UPPER MALL

12 SUSSEX HOUSE · Emery Walker's Works Est. 1886
14 SUSSEX COTTAGE · Kelmscott Press 1891-1898
15 Doves Bindery Est. 1893
16 Kelmscott Press Jan-May 1891
24 RIVER HOUSE · Cobden-Sanderson 1903-1909
26 KELMSCOTT HOUSE · William Morris 1878-1896

WILLIAM MORRIS
AND THE ART OF THE BOOK

WILLIAM MORRIS
AND THE ART OF THE BOOK

*With essays on William Morris
as Book Collector by Paul Needham,
as Calligrapher by Joseph Dunlap,
and as Typographer by John Dreyfus*

THE PIERPONT MORGAN LIBRARY
OXFORD UNIVERSITY PRESS

Oxford University Press

NEW YORK LONDON OXFORD GLASGOW
TORONTO MELBOURNE WELLINGTON CAPE TOWN
IBADAN NAIROBI DAR ES SALAAM LUSAKA ADDIS ABABA
KUALA LUMPUR SINGAPORE JAKARTA HONG KONG TOKYO
DELHI BOMBAY CALCUTTA MADRAS KARACHI
ISBN 0-19-519910-3

COPYRIGHT © 1976 BY THE PIERPONT MORGAN LIBRARY
29 EAST 36 STREET, NEW YORK, N.Y. 10016
PRINTED IN THE UNITED STATES OF AMERICA
LIBRARY OF CONGRESS CATALOGUE CARD NUMBER 76-29207

CONTENTS

Lenders	8
Preface by Charles Ryskamp	9
Memories of Collecting by John M. Crawford, Jr.	13
WILLIAM MORRIS: BOOK COLLECTOR by Paul Needham	21
WILLIAM MORRIS: CALLIGRAPHER by Joseph Dunlap	48
WILLIAM MORRIS: TYPOGRAPHER by John Dreyfus	71
CATALOGUE by Paul Needham	97
The Library of William Morris *Documentation — Manuscripts — Incunabula*	97
Calligraphy	110
Printing and Book Design *Before Kelmscott — Founding the Kelmscott Press: Types and Paper — The Press Under Way — Chaucer*	115
PLATES	I–CXIV

LENDERS

Birmingham City Museums and Art Gallery
Bodleian Library
British Library
Cambridge University Press
Edward Laurence Doheny Memorial Library
Fitzwilliam Museum
The Grolier Club
Metropolitan Museum of Art
St. Bride Printing Library
Society of Antiquaries
Victoria and Albert Museum

Sanford L. Berger
Bernard H. Breslauer
Elizabeth and J. Ben Lieberman
Paul Mellon
Earl of Oxford and Asquith
Gordon N. Ray
John A. Saks
Norman H. Strouse
Dr. Gerald Wachs

Preface

THIS VOLUME IS INTENDED TO CELEBRATE THE ACcomplishment of one of the foremost men in the history of the printing arts, William Morris. It is hard to think of anyone since the Renaissance who excelled in so many fields: poet, novelist, translator; designer and decorator of stained glass, tapestries, wallpapers, carpets, furniture; leader in social and political causes; printer, typographer, illustrator, calligrapher; and equally distinguished as a collector. His was a life devoted to many of the arts, but recorded in this volume are only his contributions to the art of the book: how he acquired his remarkable knowledge and his craftsmanship, and how he created a rebirth of fine bookmaking and private presses for England and America, and for Europe.

William Morris and the Art of the Book is one of several studies and exhibitions sponsored in recent years by The Pierpont Morgan Library to examine more fully the achievements in typographical arts, in paper and binding, type and illustration, craftsmanship and presswork. As I have written elsewhere, we are attempting to fulfill our obligation to educate and delight the public through publications about our own extensive holdings of books of magnificent beauty, with contributions of comparable quality from other collections. The Morgan Library is a museum of the book, as well as a research library and a museum for other arts. This can be seen particularly well in the case of William Morris. The Library contains the largest and finest collection of manuscripts and printed books from Morris's library. No English man of letters was a more discriminating collector, nor more concerned with design and bookmaking, than William Morris, and the volumes from his library are remarkable signs of his taste and the source of many of his ideas for typography, calligraphic forms, illustration, and binding.

From the libraries he used while a student, and then from the manuscripts he pored over in the British Museum, Morris evolved his own conceptions of calligraphic design which were firmly rooted in European medieval traditions. Later the public libraries were supplanted by his own library of medieval manuscripts and early printed books. In the present book, for the first time, there are substantial studies of Morris as a book collector and as a calligrapher. Afterwards we follow his career as a typographer, during the 1860s and 1870s before the establishment of the Kelmscott Press, and then from the end of 1889 with the creation of the new press. The Kelmscott Press years are documented with drawings and designs for typefaces and for illustrations, photographs, process blocks, early proofs, trials of type, specimens of type, and then papers, ornaments, page proofs, the books themselves—on paper and on vellum—and the special bindings for these books. Every piece of Morris's printing is not discussed, nor every one of his calligraphic manuscripts. Instead, we pursue in considerable depth the major achievements of Morris in relation to the art of the book. It is hoped that the reader through text and illustration can therefore easily follow the evolution of Morris's work to its culmination in that masterpiece, the Kelmscott *Chaucer* of 1896.

We should not have been able to do this were it not for the great interest in Morris's library and his printing on the part of the founders of the Morgan Library, Pierpont Morgan and his son J. P. Morgan, and its first Director, Miss Belle Da Costa Greene. Equally significant in building our rich collections have been the gifts of John M. Crawford, Jr. Mr. Crawford formed one of the finest collections of materials relating to Morris, and the books produced by him. This volume gives eloquent testimony to Mr. Crawford's discernment as a collector.

At the Morgan Library there are many other examples of his special purchases and his gifts which are not revealed here: they came from his own collections, or they were selected for the Library in consultation with the Director or curators, or simply chosen by Mr. Crawford. They range from the Japanese printed temple sutra of the Empress Shōtoku (770) to the bound volume of prints known as the Victories of the Emperor Chi-en Lung (1769–1774) to the portrait drawing of Diderot by Greuze. In 1957 he gave the Library the important calligraphic manuscript by Morris of Kormák's Saga. In the last year or two he has given us a manuscript of the Bible, written in England or France about 1280, whose provenance can be traced without break from a Benedictine monk of St. Augustine's monastery, Canterbury, about 1300 named Nicolaus de Bello. There were also marvellous examples of modern fine printing: a copy of the Bruce Rogers Oxford Lectern Bible bound by Roger Powell and one of twelve copies on vellum of The Golden Cockerel Press *Four Gospels* with decorations by Eric Gill.

Along with the many items in his Morris collection which Mr. Crawford recently gave to the Morgan Library, which are mentioned in this book, are other notable volumes, prints, drawings, and manuscripts which are not included: an extensive group of Kelmscott ephemera—order forms, prospectuses, trade notices, certificates, letterheads, etc.; many proof leaves; letters of Morris, the Morris family, and Burne-Jones; a sketchbook of May Morris; a Burne-Jones sketchbook (probably 1877), his pencil drawings of "The Fiery Furnace" and "The Iron Chariots" (apparently for a projected Kelmscott edition of the Bible), a "Study for Cupid" (1880), and his "Study for Flora"—a finished colored tapestry design made with William Morris; Morris's autograph manuscript of "Gossip about an Old House," which is being specially reprinted in conjunction with the publication of this book; eighty-nine colored designs for stained glass windows from Morris and Co., some of them initialled by Morris; vellum copies of the Kelmscott editions of *The Order of Chivalry* and Swinburne's *Atalanta in Calydon*; and, from Morris's library, a French thirteenth-century Bible manuscript of the four books of Kings.

Mr. Crawford is today known most widely as a collector of Chinese art, especially of calligraphy and painting. This book will show he was equally distinguished as a collector of William Morris. In both Eastern and Western arts, Mr. Crawford has a superb eye for the shaping of forms of letters or characters and designs in black on white. His own contributions to collecting in our time are so considerable that we have asked him to tell his own story in an introductory essay to this book. We are deeply grateful to Mr. Crawford for his encouragement and his discrimination in the arts, especially in the art of the book, and in his generous sharing of his collections through individual loans, in entire exhibitions, and in his gifts to many institutions, particularly to The Pierpont Morgan Library. His was the inspiration for this book and for the exhibition held at the Morgan Library, 7 September to 28 November 1976.

Paul Needham, Curator of Printed Books and Bindings in the Morgan Library, has been chiefly responsible for the selection and the organization of the material in this book; he has also written one of the three principal essays and the catalogue proper. Joseph Dunlap and John Dreyfus joined him to write the essays on "William Morris: Calligrapher" and "William Morris: Typographer." We are above all indebted to them. Three scholars of William Morris were extremely generous in sharing their knowledge with us: David Donaldson, Professor Norman Kelvin, and, again, Joseph Dunlap. Sanford Berger and Nicholas Poole-Wilson were also very helpful, as were the following members of the staff of the Morgan Library: Charles V. Passela, who took most of the photographs, John Plummer, William Voelkle, and Mrs. Susan E. Vosk. The many institutions and private collectors who have made possible descriptions and reproductions of Morris's works, and loans for the exhibition, are listed separately; we are most grateful to all of them, and to The Charles W. Engelhard Foundation for generous support of this and other publications of the Morgan Library. Our friends at The Meriden Gravure Company and The Stinehour Press have worked closely with us to create a book worthy of the standards of design and printing established by William Morris.

CHARLES RYSKAMP
Director

Memories of Collecting

COLLECTING HAS BEEN THE PRINCIPAL THEME OF MY life. In my childhood our family traveled extensively in the United States and Europe and spent much time in museums, churches, cathedrals, châteaux, and palaces. These experiences made indelible impressions on me and European history became a passion. While still very young, I devoted many hours to reading volumes of richly illustrated histories of the world; the accounts of the various Asiatic empires had a particular fascination for me. History continued to be my favorite subject in school.

A faculty wife at Brown University, Celia Sachs Robinson (now Mrs. Richard Stillwell), recently recalled my talking to her in 1936 about collecting, and insisting as well that a perceptive collector is, in his own way, a fine artist. During these university years I subscribed to the Limited Editions Club for a year and its handsome volumes gave me great pleasure, and later at the Harvard School of Education in 1938–39, I responded to an advertisement for the Limited Editions Club Book Exchange. While on a visit to New York in February 1939, I stopped in to see the bookseller Philip C. Duschnes and we became friends. On subsequent visits, Phil aroused my interest by showing me outstanding examples of modern fine printing. Within a few months I had bought a copy of Bruce Rogers's Oxford Lectern Bible and a Kelmscott Chaucer: a good start. Discussing books led naturally to reading books about books, and by the time I came to New York to live in September 1941, I was an ardent collector of modern fine printing. The postwar years brought expanding collecting interests in medieval manuscripts, early and later examples of printing, fine bindings, unusual rarities. I also acquired English calligraphic manuscripts by William Morris, Edward Johnston, Graily Hewitt, and others.

At this time, I became increasingly fascinated by the protean figure of William Morris, and particularly by the extraordinary range of his achievements. I added steadily to my Morris collection, soon finding it necessary to specialize largely (though never exclusively) in the Kelmscott Press. Especially memorable was the 1943 acquisition of my first unique material: a portfolio of most of the proofs for the Burne-Jones illustrations of the Kelmscott Chaucer, with detailed corrections by the artist. At the Sotheby sale in 1956 of Sir Sydney Cockerell's Kelmscott Press material I was able to acquire many unique items which had never previously been on the market. Upon the occasion of the meeting of the Bibliographical Society of America at Brown University in October 1959, there was an exhibition of my Morris material at the University's John Hay and Ann Mary Brown Libraries. The following year the printed catalogue was sent to the Bibliographical Society membership. Just prior to The Grolier Club exhibition of this material in the autumn of 1964, many more important items from the Cockerell sale came to me when Norman Strouse and I made an exchange of his unique Kelmscott material for my best Doves Press pieces. It was a happy exchange on both sides, and we became better friends than ever.

During these same years I found my collecting instincts aroused by a second and equal passion: Oriental art. The climax of extensive travel in my youth was a long trip around the world during the later months of 1937 and early 1938. April found me in Japan where Chinese and Japanese art had a profound impact on me. Later, my interest in and knowledge of this area was increased by reading and by my visits to the Far Eastern sections of many American museums. As the Second World War was ending, I began collecting Chinese art, concentrating for the first few years on Chinese porcelains of the seventeenth and eighteenth centuries.

Early in 1951 I met the knowledgeable dealer Joseph U. Seo. As a result, my collecting broadened into scholar jades, large sculpture, small gilt bronzes, Shang bronzes, and some Japanese art. One hot afternoon in August 1955 marked a turning point for my collection: I acquired several Northern Sung landscape paintings, followed shortly by Sung calligraphies. I feel sure that my previous involvement with the fine points of Western printing types and their artistic uses helped me all the more to appreciate the limitless subtleties of Chinese calligraphy.

It was perfectly natural to collect the twin arts of Chinese painting and calligraphy, for both arts use the same materials (ink, ink stones, paper or silk, and brushes) and both are essentially based on the individual brushstroke. They are the heart and soul of Chinese civilization—the oldest and most continuous our world knows—because they express the essence of Chinese culture. Chinese connoisseurs have always admired them, together with poetry, above all other arts: poetry, painting, and calligraphy are the 'three perfections.'

As the years passed, my collection broadened to include examples of both arts from the late Tang (ninth century) through the eighteenth century in every format—hanging scrolls, hand scrolls, album leaves, fans, and artists' letters. More recently, the distinguished connoisseur-dealers Alice Boney and Jean-Pierre DuBosc have been very helpful to me. By the early 1960s, the pursuit of these arts became so absorbing that I stopped my other collecting, believing that I was making an almost unique contribution to collections in the Western world.

After thirty-five years in New York City, I know I could not have led such a vital and varied life elsewhere. I agree with Zubin Mehta, the next music director of the New York Philharmonic Orchestra, that culturally 'New York is the place to be. It's the center of the world, the lion's den.' My involvement in my favorite local institutions—The Grolier Club, The Pierpont Morgan Library, the China Institute, the Asia House Gallery, and the Metropolitan Museum—has brought me many wonderful friends. New York attracts creative people from all over the world; they make it the unusual cultural center it is. Frequent attendance at art exhibitions and outstanding performances of music, dance, and the theatre have been inspiring experiences.

The first outstanding bookman I came to know in New York was Bruce Rogers, the great American book designer, who taught me much about typographical distinction. In early 1947 Phil Duschnes and I revived the *Colophon*, a notable publication of the 'Thirties, as *The New Colophon*, with the same editors, Elmer Adler, John T. Winterich, and Frederick B. Adams, Jr. Each became a good friend and I learned much from these remarkable men. Most of our editorial meetings were delightful luncheons at the Drake Hotel, but a few were held at the Morgan Library after Fred Adams became its Director in December 1948. These years of *The New Colophon* were most rewarding but, regrettably, contemporary conditions made its continuance impossible, although Winterich felt that the contents of *The New Colophon* were superior to those of the *Colophon*.

The New Colophon meetings led to my acquaintance with and deep admiration for the Morgan Library. In 1949 I was among the original group invited to become a Fellow of the Library: that early group is sadly dwindling now. Inspired by the Morgan, I became interested in old master drawings and even started collecting in a small way; my chief acquisition was a fine G. B. Tiepolo. However, as I have said, Chinese painting eventually took over. As time passed, I gave the Library a wide variety of gifts, including several drawings and William Morris's calligraphic manuscript of *The Story of Kormak,* recently published by the William Morris Society in England. In 1959 I was honored to become a Life Fellow and now in 1976 a Fellow in Perpetuity. It has been a pleasure to serve for many years as a member of the Library's informal Music Committee, so ably led by Alice Tully. I have vivid memories of many outstanding exhibitions at the Morgan, and another great pleasure has been my friendships with the Library's gifted staff.

In the spring of 1958 Fred Adams accepted and encouraged my hesitant suggestion for an exhibition at the Morgan Library of my rapidly growing collection of Chinese paintings and calligraphies, with which he was familiar. What a bold decision it was to agree to exhibit a collection of two then little-known arts, especially as they had no connection with the Library's own collecting interests. I recommended that Laurence Sickman, Director of the William Rockhill Nelson Gallery of Art and Mary Atkins Museum in Kansas City, be the editor of the catalogue we planned for the exhibition. I had been to his wonderful museum several times and had read his superb history of Chinese sculpture and painting. In preparing for the exhibition, Laurence Sickman's first visit to me was memorable. After examining several paintings, he agreed to become editor as well as one of the authors of the catalogue. He was the first important Chinese art scholar and connoisseur to be enthusiastic about the collection.

With the assurance of such outstanding scholarship, Fred Adams and I agreed that the catalogue should embody our shared ideals of what a well-designed and finely printed book should be. It was three years in the making, but it surpassed our expectations. There was never any question as to the designer and printer of the text and of the many collotype illustrations. The long, close collaboration of Fred Adams, Laurence Sickman, and myself with Joseph Blumenthal of The Spiral Press and Jim Barnett and Harold Hugo at The Meriden Gravure Company made a prize-winning book possible. Working with these close friends was a truly remarkable creative experience. While in London in 1963, I commissioned Roger Powell and Peter and Sheila Waters to design and execute two bindings for this catalogue, one inspired by Chinese painting and the other by Chinese calligraphy. The results are two brilliant bindings.

The exhibition reopened the Morgan Library in October 1962 after its partial closing for three years for reconstruction and additions. Seeing my entire collection exhibited together for the first time (at home, of course, the scrolls are kept rolled up in drawers), was a deeply emotional experience for me. The entrance hall, the large exhibition hall, and the cloister were all used. The four gala openings at the Morgan, followed by a fascinating lecture series, launched the collection publicly in great style. Later, the exhibition moved on to the Fogg Museum of Art and, finally, to the Nelson Gallery. At the suggestion of Michael Sullivan, then Professor of Chinese Art at the University of London and now at Stanford University, Fred Adams arranged for a somewhat smaller version of the exhibition to be sponsored during 1965 and 1966 by the Arts Council of Great Britain at the Victoria and Albert Museum, the National Museum in Stockholm, and the Musée Cernuschi in Paris. In recent years I have lent to many exhibitions in New York and elsewhere. The 'Chinese Calligraphy' exhibition suggested and organized in 1970 by Jean Lee for the Philadelphia Museum of Art with the catalogue by Tseng Yo-ho Ecke, the first exhibition in the West devoted solely to this subject, was particularly notable for all of us. Afterwards it traveled to the Nelson Gallery and the Metropolitan Museum of Art, New York. Later there was 'Friends of Wen Cheng-ming. A View from the Crawford Collection,' which was originally shown at the China Institute in New York, starting in October 1974, continuing on to the Nelson Gallery, the Seattle Museum of Art, and the Bell Gallery at Brown University in 1975.

What is the purpose of exhibitions? Most collectors have something of the missionary in them. What they collect has so much meaning for them that they wish to share the pleasures and spread knowledge of their particular field. Properly selected and organized, I believe that public exhibitions can be important cultural landmarks. Catalogues often record new knowledge and insights. In my experience, still another way of sharing is to receive scholars, students, and others for intimate viewing sessions of my collection. A few people examining a gradually unrolled scroll on a table is an extraordinarily rewarding aesthetic experience.

Through the years I have enjoyed giving books and manuscripts to various libraries and works of art to museums. I now take particular pleasure in donating my cherished William Morris collection to the Morgan Library. The exhibition itself is the culmination of an idea I had years ago to show Morris as a man of books, my inspiration being based on the fact that the Morgan Library has owned since its inception much of Morris's own collection of medieval manuscripts and early printed books, which he gathered together so joyously in his late years. Further, the Morgan has acquired over the years many important Kelmscott Press items.

Under the enthusiastic and discerning directorship of Charles Ryskamp, I have found it a pleasure to work with John Dreyfus, the English typographer and typographic historian, Joseph Dunlap, American Secretary of the William Morris Society, and Paul Needham, Curator of Printed Books at the Morgan Library, in the creation of this catalogue. The design and printing by The Stinehour Press and the illustrations prepared by The Meriden Gravure Company have guaranteed a presentation of the highest quality. My thanks goes to each of them but most especially to William Morris.

JOHN M. CRAWFORD, JR.

If I were asked to say what is at once the most important production of Art and the thing most to be longed for, I should answer, A beautiful House; and if I were further asked to name the production next in importance and the thing next to be longed for, I should answer, A beautiful Book. To enjoy good houses and good books in self respect and decent comfort, seems to me to be the pleasurable end towards which all societies of human beings ought now to struggle.

WILLIAM MORRIS
Some Thoughts on the Ornamented MSS. of the Middle Ages

William Morris: Book Collector

A MAN OF LETTERS, EVEN WITH A STRONG DISTASTE FOR book collecting in the ordinary sense, will yet inevitably form a library of one kind or another, and it is natural that we should be curious as to how the man is reflected in his books. A perversion of this natural curiosity led William Henry Ireland, in midst of his more audacious forgeries, to draw up a pseudo-catalogue of Shakespeare's library, and to inscribe false Shakespearean annotations in copies of Elizabethan books which Shakespeare was generally believed to have read. Less deluded studies have been made of the libraries of Donne, Ben Jonson, Swift, Walpole, and Gibbon among many others. There have been investigations of an author's annotations of his books (Gabriel Harvey, Macaulay), or even of a single book (Keats's Shakespeare). Most recently a multivolumed set of facsimiles has been issued of the library sale catalogues of English poets and men of letters. It is strange, however, that nothing of substance has been written on the library of William Morris, for his book collecting is richly documented and greatly revealing. The longest connected account we have of Morris's library is the dozen perceptive pages contributed by May Morris, with the help of Sydney Cockerell, to vol. XXIV of the *Collected Works* of her father; but good as this is, it is a fragmentary story.

More, probably, than any other English writer Morris cared, and cared intelligently, about the appearance of his publications. This concern, attested throughout his writing career but reaching fruition only in the last years of his life, is accurately mirrored in his library. Indeed we can go farther and assert that Morris's distinctive taste in book design was to a great extent engendered and confirmed by his own adventures in collecting. Essentially Morris wanted in his library books which pleased him by their beauty, and he was fortunate, particularly in his later years, in being financially able to acquire them. If we exclude men such as William Beckford and Robert Curzon, who are known primarily as collectors rather than authors, it may be claimed that William Morris possessed a library of higher quality than any other major English literary figure. The one potential competitor who comes immediately to mind, his friend and mentor John Ruskin, is no sooner named than disqualified, for Ruskin had a barbaric way with books, and they often left his library in much sadder shape than when they entered it. He wrote disfiguring notes in some of his medieval manuscripts; from many others he cut out leaves and miniatures and gave them away, or rearranged them in albums; his library shelves were a Procrustean bed, and books which were too tall to fit in them were sawed down to size. In short, Ruskin's name figures prominently in the annals of enemies of books. And in any case, though he owned a small group of medieval manuscripts of exceptional quality, on balance Ruskin's library was neither as closely integrated nor as valuable as Morris's.

William Morris's initiation into the world of antiquarian books came at about the age of thirty through the bookseller Frederick Startridge Ellis (1830–1901). Ellis was a self-taught man of broad literary interests, who from an early age worked as apprentice and assistant in several London bookshops. In 1860 he set up on his own in King Street, Covent Garden, where he built an important stock and a distinguished clientele, including John Ruskin and Dante Gabriel Rossetti. Besides his bookselling Ellis later also published on a small scale for his literary friends. Morris was brought into F. S. Ellis's shop by Swinburne, about 1864. Ellis later recalled that Morris's first acquisition of any significance was a fine copy of the Ulm edition of Boccaccio, *De claris mulieribus*, 1473. The price of this great illustrated incunable seemed high to one as yet unacquainted with the market in rare books and manuscripts—some £26—and Morris returned with his friend Edward Burne-Jones to inspect the volume once more before deciding to take it. Indeed the price was high by some standards, for this was more money than a housemaid might earn in a year. The Boccaccio became familiarly known to Morris and Burne-Jones as the Yellow Book, on account of its stained-vellum binding. Among other books Morris bought about this time were a copy of the Nuremberg Chronicle (Latin text, 12 July 1493), and the first Latin edition of Sebastian Brant's *Narrenschiff* (Basel, 1497), reusing the woodcuts of the original 1494 edition.

Morris's meeting with F. S. Ellis was significant in more ways than one, for the two men became firm friends, and their lives ran closely together after this. At the time they first met, Morris's only separate publication had been the 1858 *Defence of Guenevere*. In 1868, with the first volume of *The Earthly Paradise*, Ellis began publishing Morris's books and he continued as his publisher for some years. Starting in 1874, following the unsuccessful experiment with Rossetti, Morris shared the tenancy of Kelmscott Manor with Ellis.

Over the next dozen years Morris continued to acquire old books and an occasional medieval or humanistic manuscript, but not, as it would seem, in any major way. An early priced catalogue of the library has survived, partly in Morris's autograph. Though undated, a flyleaf note by Cockerell assigns it to 'about the year 1876.' This date is at least plausible; the most recent volume entered is a publication of 1875. This catalogue enumerates 291 printed books, the last 54 of which comprised Morris's Icelandic collection, which was still growing. The Icelandic group was for the most part a working collection of standard editions of the late eighteenth and nineteenth centuries, but there were a few seventeenth-century imprints, including four of the Sagas printed at Skálholt in Iceland, 1688–90. The non-Icelandic books were generally pre-1800, but here too a number of nineteenth-century reprints of medieval literature were included.

The bibliographic detail in the catalogue is scanty, generally extending to no more than the place and date of issue, and a note when the books contain woodcuts. The many scribal errors suggest that the catalogue was put together rapidly, in part by dictation. Except for the Islandica there is no clear subject division, although broadly speaking the first entries are of earlier books. The majority of the entries fall into three main categories: incunabula, sixteenth-century continental books, and sixteenth- and seventeenth-century English books. There are only thirteen incunabula, and it is strange that the Ulm Boccaccio is not among them; the Nuremberg Chronicle is also not entered, but that is because Morris gave it to William Michael and Lucy Rossetti in 1874 as a wedding present. There are about seventy continental imprints of the sixteenth century, at least twenty-eight of which date to 1501–1520; and close to sixty English books before 1700.

The small group of incunabula includes some important woodcut books: the Latin *Narrenschiff*, already mentioned; the *Schatzbehalter* (Nuremberg, 1491), its date carelessly entered as '1590?'; and the *Mer des histoires* (Paris, 1488–89). This last, however, Morris noted as 'not in good condition,' with two leaves missing from the first volume, and repairs in the second. The edition of Horace listed as Strassburg, 1598, must surely be a careless reference to Grüninger's 1498 edition, and this too is a handsome, profusely illustrated work. Two other important illustrated incunabula in the library were the Bergomensis, *De claris mulieribus*, and Hieronymus, *Vita ed epistole*, both printed by Laurentius de Rubeis in Ferrara, 1497. The Bergomensis was a bargain: Morris paid £2 2s for it but, as he noted, the book was 'Worth more than 2 guin. worth 7 or 8 guin.'

The most expensive of the incunabula was not illustrated: the first edition of Wolfram von Eschenbach's *Titurel* (Strassburg, 1477), which Morris noted as costing £10. In fact, he probably did not buy the book, as this is apparently the copy which F. S. Ellis presented to him, and which the Morgan Library disposed of as a duplicate in 1971. The prices assigned to the incunabula total just under £48, or an average of £3 13s a volume; if the *Titurel* is excluded, the average is £3 3s. In comparison the standard price of three-decker novels, long maintained at an artificially high level, was 31s 6d, or just half what Morris paid for his fifteenth-century books.

The sixteenth-century continental books seem to have been collected as casually as the incunabula, though here too the general preference was naturally for woodcut illustrations. Morris's most expensive purchase, at £30, was a fine copy of the 1533 Paris edition of *Lancelot du Lac*; his only other purchase at this level was the 1551 Giunta *Epistole et evangelii* with numerous woodcuts, £25. The famous *Hypnerotomachia Poliphili* was present in the second edition, Venice, 1545 (with the exception of a few recuttings its woodcuts are repeated from the 1499 Aldine edition). Three important Dürer publications are recorded: the *Epitome in diuae Mariae historiam*, 1511 (£2 10s), the *Vnderweysung der Messung*, 1525 (£1), and the *Apocalypsis cum figuris*, edition not identified (£10). Another great Nuremberg woodcut book was Ulrich Pinder's *Rosenkrantz Mariae*, 1505, but the second half only. Morris's active involvement with textiles and dyeing must have encouraged him to acquire G. V. Roseto's *Plictho. De larte tentori* (Venice, 1540; £25, the Yemeniz copy). And in view of Morris's interest in calligraphy during these years, one more volume must be mentioned, a collection of four rare early Italian writing books: Arrighi's *Operina* (Rome, 1522) and *Modo de temperare le penne*, 1523; the 28-leaf edition of Tagliente's *Vera arte delo excellente scriuere*, 1525; and Ugo da Carpi's 1525 collection of several writing masters, *Thesauro de scrittori*, lacking a leaf. The four tracts, bound up together, cost Morris two guineas.

The English books do not require extensive notice. For the most part they are chronicles and histories: Hall, Fabyan, Holinshed, and Elizabethan translations of Livy, Plutarch, Sallust, Thucydides, and others. Considering the prices he paid, Morris must have looked on many of these volumes as falling somewhere between plain reading copies and antiquarian purchases. His copy of the Philemon Holland translation of Pliny's Natural History, for instance, the 1601 first edition, cost only £1 5s, as against the £5 paid for the 1817 Southey reprint of the *Morte d'Arthur*. Given his strong dislike for nineteenth-century printing in general, it was natural for him to buy older books cheaply. From occasional notes in the catalogue it is evident that not all the books were in fine condition. There are a number of references to missing leaves and other imperfections, and against one volume, a 1568 edition of Cicero's Offices, Morris wrote succinctly, 'grubby little book.'

What is perhaps most striking about this early state of Morris's library, as compared with its makeup in the 1890s, is the scarcity of medieval manuscripts. On the first page of the catalogue Morris listed only six manuscripts, most of which he later disposed of. Two are Italian humanist works: a Latin version of Xenophon, *De legibus Lacedaemoniorum*, and a handsome *Astrologia* of Christianus Prolianus, closely related to the first printed edition of the text, Naples, 1477. The other manuscripts cannot be precisely identified: a Bible assigned to Italy, thirteenth century, had 'many of the letters cut out,' whereas a fourteenth-century French Bible for which Morris paid £16 was in 'good condition.' Morris did not yet look on himself as a serious collector of manuscripts. Following these early manuscripts, and with consecutive numbering, Morris listed seven of his own calligraphic manuscripts.

This first catalogue reveals Morris as a lover of books, with a not unexpected taste for early printing and woodcut illustration. But it would be excessive, especially in light of the opportunities available at the time, to call him a bibliophile. His library contained many good early books but also many which were not particularly good. The average figure Morris paid for these books and manuscripts, exclusive of the Icelandic section, was just under £2 9s; if the nine volumes for which he paid above £10 apiece be excluded, the average falls to £1 16s. This is no more than a gentleman of the time might lay out for a standard nineteenth-century collection of travel, sport, and literature. Morris's acquisitiveness was still well under control.

Very little is known about William Morris's book collecting in the 1880s, until the end of the decade. These were the years when the main force of his restless energy was directed toward socialist activities, and it is tempting to think that at least for part of this time the collecting of old books had in consequence a diminished appeal. One piece of evidence, a letter by Morris, would seem to support this view, except for a problem concerning its true date. In January 1883, Morris enrolled formally in the radical Democratic Federation. But already 'in the previous October,' according to his biographer J. W. Mackail, Morris 'had sold the greater part of his valuable library, in order to devote the proceeds to the furtherance of Socialism.' Mackail's knowledge of this episode was based on a letter from Morris to F. S. Ellis, whose firm, Ellis & White, purchased the books. The full text of the letter has not survived, but Mackail cites specifically the Ulm Boccaccio as one of the volumes sold and then quotes Morris as follows:

> If the modern books are unsaleable, perhaps you would let me take them out after your valuation, as I have no idea what they are worth to sell (though beastly dear to buy), and though I hate them and should be glad to be rid of them as far as pleasure is concerned, they are of some use to me professionally —though by the way I am not a professional man, but a tradesman.

These words have a vigorous sound, and are consistent with the notion, repeated by most of Morris's biographers, that a fairly clean sweep was being made. But two problems are involved. The first is that Mackail's date of the letter is almost certainly incorrect; it must have been written in October 1880 rather than October 1882. On 6 October 1880 Ellis's partner David White wrote to Morris, saying, 'I have gone over the books received from you yesterday & value them at £130.' He went on to mention several titles which had been on a list provided by Morris, but had not been delivered. These included the aforementioned Icelandic books from the seventeenth-century Skálholt press, the 1497 Ferrara edition of Bergomensis, and a modern work, Gailhabaud's *L'Architecture*, 1858. It is impossible to avoid the conclusion that this is the very sale of books that Mackail assigned to 1882, and associated with Morris's growing interest in socialism. Mackail even mentions these same Skálholt imprints as among those supposedly sold by Morris at the later time. The most likely explanation for the error is that Morris's letter lacked the year date (as do many), and that Ellis supplied it out of a faulty memory. It is curious, however, that Ellis seems to have misremembered also the motive for the sale; he went so far as to speak of Morris's having 'sacrificed' his books 'at the altar of Socialism.' It is worth recalling in this context that May Morris, who was eighteen at the time, later laid the cause more vaguely but probably more accurately to 'the cares of a house and of business.'

The second problem with Mackail's, and through him the general, account is that by no means was 'the greater part' of the library sold. Though it is difficult to say precisely how much Morris did dispose of, the great majority of his books, including the early books, were retained. This may be shown in two ways. David White valued the books Morris was selling at £130. Since the total value of the contents of the 1876 catalogue was about £600, it would appear that Morris was selling at most about twenty percent of his library.

A comparison of the entries in the 1876 catalogue with those of a catalogue of the library dating to 1890–91 leads to much the same conclusion (though what we call here the 1876 catalogue is undated, it is certain that it precedes the sale of books in 1880). Thirteen incunabula, it will be recalled, were entered in the 1876 catalogue; of these, nine appear again in the catalogue of 1890–91. The presumption is natural that these are the same copies, and in two instances we may be quite certain that they are: the *Mer des histoires* is described with the same faults in both catalogues; and the Regiomontanus *Epytoma in Almagestum* (Venice, 1496) is in both catalogues noted as bound with two other mathematical treatises. Sebastian Brant's *Stultifera navis* does not appear in the 1890 catalogue, but the Pierpont Morgan copy of this book was in Morris's library at his death; and since it bears Morris's signature as from Red House, it must have been acquired before he moved from there in late 1865. This leaves open the possibility, a slim one, that he sold this book in 1880, and purchased the same copy again sometime after 1890. The two Ferrarese books mentioned earlier also do not appear in the 1890–91 catalogue, although Morris did own copies of both books in the 1890s. These are now in the Morgan Library, but they contain no clues as to when Morris acquired them.

Of the nine sixteenth-century books mentioned specifically by title above, six are listed in the 1890–91 catalogue, and two others—the 1533 *Lancelot* and the Dürer *Epitome*—were in any case in Morris's library at his death. A random sample of twenty-five sixteenth-century imprints from the 1876 catalogue shows that eighteen reappear in the 1890–91 catalogue. Another three are known to have been in Morris's library at his death, it being uncertain whether these were his original pre-1876 acquisitions or later purchases. One at least may well have been the same copy: a Venice 1533 Ovid is described in the 1876 catalogue as a 'poor copy,' and the copy in the 1898 sale of Morris's books lacked the titlepage.

Morris later regretted having surrendered two books in particular, the Ulm Boccaccio and the *Schatzbehalter*, both masterpieces of early German book illustration; he eventually reacquired fine copies of both. What else did Morris dispose of at this time? A French Josephus (Paris, 1530) seems to have disappeared from the library, as did a copy of Boccaccio, *De la généalogie des dieux* (Paris, 1531) for which Morris had paid £10, and the 1512 Antwerp edition of Ludolphus de Saxonia's Life of Christ, with woodcuts repeated from the 1503 Antwerp edition. The Dürer Apocalypse, which could have belonged to any of several editions, also drops from sight. Most of Morris's illuminated manuscripts were very likely now sold. The Italian Bible with missing initials appears again in the 1890–91 catalogue as also, apparently, the French Bible. But the four other manuscripts are absent. One of them, the Prolianus, *Astrologia*, was certainly sold in the early 1880s. It was shortly afterward acquired from Ellis & White by the Earl of Crawford, and is now in the Rylands Library. The Ulm Boccaccio was purchased by another steady customer of Ellis & White, Henry Huth. After the first Huth sale in 1911 it came into the possession of Morris's old friend Charles Fairfax Murray, and was one of his generous benefactions to the University Library, Cambridge. There may also have been other books acquired by Morris between 1876 and 1880 and then sold, leaving no trace in his catalogues. But despite the loss of some important volumes, there was no major discontinuity between Morris's pre-1880 and post-1880 library. There was still material here to interest the amateur of early printing and book illustration.

One such amateur, the crossing of whose path with Morris's led to new directions in the lives of both men, was Emery Walker, a Hammersmith neighbor. Walker, like Ellis, was a self-taught man who had gone to work at an early age. He became the co-proprietor of a firm doing high-quality process engraving, and the special strength of his lifelong love of fine typography and book design was his thorough practical knowledge of the printing trade. Morris and Walker first became acquainted, about 1884, at a Socialist meeting in Bethnal Green, but Walker was already known by sight to the inhabitants of Kelmscott House as the 'brown velveteen artist.' May Morris has quaintly described how he could be seen 'in his hurried flittings past the windows on the way to business, sometimes leading by the hand a pretty little maid in white muslin.'

Morris invited Walker to visit the Kelmscott House library, and to explain, from his practical knowledge of printing, 'certain details of the methods of early printers, a subject which was beginning to occupy his mind.' Thereafter Walker became a frequent, almost daily visitor; he was drawn into the activities of Morris's other enthusiasm, the Society for the Preservation of Ancient Buildings (S.P.A.B.), and, later, both men were active in the Arts and Crafts Exhibition Society.

At the first Arts and Crafts exhibition, in November 1888, Walker was persuaded by T. J. Cobden-Sanderson to deliver a lecture on 'Letterpress Printing and Illustration,' in the preparation of which he consulted with Morris. Walker's talk was centered around a series of lantern slides showing enlargements from early printed books and related manuscripts. Oscar Wilde covered the lecture for the *Pall Mall Gazette*, and described the audience's enthusiastic reception of the slides. Some of these were almost certainly taken from Morris's own books, and his tastes are evident in the choice of others, the Ulm Boccaccio for instance. The slide of Vicentino Arrighi's writing book was probably taken from Morris's copy; it 'was greeted,' Wilde tells us, 'with a spontaneous round of applause.' The illustration of the *Mer des histoires* was also probably taken from Morris's copy, as likewise the 1556 Basel Polydore Vergil, with Holbein borders.

The sight of these enlargements, particularly of the fifteenth-century types, made a great impact on Morris. He returned home with Walker and proposed that night that they design a new fount of type. There was a new craft to be mastered and, his purpose set, Morris entered on the enterprise with his customary vigor and decisiveness. As a consequence, according to Walker, Morris now 'supplemented his collection of Early Printed books by buying examples of every fine type—gothic or roman—obtainable.' May Morris also noted a change in the tempo of book collecting at this time, recalling that in 1888 'his taste of younger days for early printed books had been developing into a practical interest in all the details of fine printing; he had also begun collecting early printed books and manuscripts, and was constantly "talking shop" with Emery Walker.'

The 1890–91 catalogue of the Kelmscott House library is for the larger part a chaotic compilation, yet it contains, when carefully used, some traces and clues to this renewed surge of collecting. The catalogue is not dated, but a flyleaf note by Sydney Cockerell states that it 'was begun in 1890 by [Morris's] daughter Jenny and was continued in the same year and in 1891 (ff. 72–91) by Morris himself.' Cockerell was soon after closely involved in the library, and his dating may be accepted with some confidence. Several books published in 1890 are listed. The catalogue includes almost 1100 volumes, though its numeration, once Morris's hand enters in, goes seriously awry. The major portion, in Jenny's hand, contains almost no bibliographical detail except for a few brief notes.

Whereas in the 1876 catalogue the manuscripts were separately treated and the early printing tended to come first with the Islandica at the end, in the 1890 version an almost random order seems to obtain. No distinction was made between early and modern books. There are occasional clusters of related works, but a 1520 Parisian Book of Hours is found next to the Hon. Wm. Herbert's *Helga, a Poem in Seven Cantos* (1815), Spenser's *Poems* (1611) is beside an 1870 *Églises de la Terre Sainte*, and W. Baxter's *British Phaenogamous Botany* in four volumes is supported at either end by a 1493 Herodian and a 1583 *Ogier le Danois*. It would be hard to explain this except by supposing that the books were shelved in the same disorder. Mackail wrote, though not specifically about this time, that apart from the early printed books Morris's library 'was not large, nor was it either carefully selected or kept with any special care. He lost books which were not precious in themselves almost as fast as he read.' This is not the place to discuss Morris's favorite reading, but we may note in passing that some of his best-loved authors are here in force, as they were not in the more selective 1876 catalogue: collected editions of Dickens and Sir Walter Scott, and almost all the novels, in French, of Alexandre Dumas.

From this disorder, it may be possible to sort out at least some early printing acquired after 1876 but before 1890, when Morris became a much more active buyer. All but one of the presumed pre-1876 incunabula are scattered in the first half of the catalogue. About six more incunabula are also entered in this position of relative exile, and these are the ones likeliest to have been earlier, isolated acquisitions. Only two of these books are well known generally: the Nuremberg Chronicle, which would have replaced the copy given to the W. M. Rossettis, and Jenson's 1476 edition of Pliny's Natural History, in Italian. This very copy was later used by Walker and Cobden-Sanderson as a model when the Doves type was being designed. There is also a Strassburg incunable of no great interest, a 1471 Sweynheym and Pannartz imprint, and a 1470 Franciscus de Retza, *Comestorium vitiorum*, noted as the 'first book printed in Nürnberg'; certainly it is one of the first several. The Nuremberg Chronicle is the only one of these of interest for its illustration. Somewhat farther along in the catalogue, but still coming before the main run of early printing, are four more Venetian imprints which were probably acquired for their typographic interest. The most significant of these was the 1476 Jacobus de Rubeis edition of Aretino's *Historia Fiorentina*. This is the book whose type served as closest model for the Golden Type, which Morris was now in process of designing.

Bound in with the catalogue is an invoice from Bernard Quaritch dated 31 March

1890 which deserves examination, as it reflects book-buying of a more committed kind than we have found previously. Ten early printed books are itemized along with a few modern ones, totaling close to £115. Some of these were of great importance: Gerard Leeu's 1480 *Dialogus creaturarum*, one of the finest of all Dutch illustrated incunables; Vérard's 1493 illustrated edition of the *Arbre des batailles* (the most expensive of the lot at £18); and a 1498 Augsburg German Aesop which, though this copy lacked four leaves, is an extremely rare edition. Jenny Morris entered most of these new acquisitions in two distinct clusters, as nos. 696–702 and 857–60, and after this point it becomes impossible even to guess at which books might have been acquired before 1889. But the sixteenth-century continental books and pre-1700 English books follow a pattern similar to that of the incunabula in that virtually all the apparent carryovers from the 1876 catalogue are scattered in the first half. The March 1890 books were not Morris's only large purchase at this time. In May he bought another batch for £200 from a group Quaritch had sent him on approval.

William Morris took over the catalogue from his daughter at fo. 74, and the remaining entries, about 150 in number, are almost all of early printing: 79 of the 140 incunabula are listed here. Most of these were probably current acquisitions, entered as they came in, for some half-dozen volumes whose dates of purchase we happen to know seem to have been recorded sequentially. Thus no. 980 is an acquisition of 1 July 1890; two acquisitions of 21 July are entered as nos. 985 and 986, and an acquisition of 13 August is no. 991. An acquisition of 15 January 1891 comes later on at no. 1051, which is about where it should be, postulating a steady rate of purchases in the interval. It becomes apparent that the library was now growing rapidly. The fact that Morris himself began entering the books also probably indicates a greater seriousness. His entries are fuller than Jenny's, and he became meticulous about copying out the colophons.

The quality of these 1890 and 1891 acquisitions was high. Morris's oldest friend among incunables, the Ulm Boccaccio, returned to his library in July 1890. His first copy had cost £26 back in the 1860s; this second one, from Quaritch, also large and fresh, was priced at £68. Three weeks later, however, Morris did pick up one remarkable bargain from Henry Sotheran: a flawless copy in contemporary binding of Gunther Zainer's edition of Rodericus Zamorensis, *Spiegel des menschlichen Lebens*, costing only two guineas. Another rare book from Ulm acquired at this time was Johann Zainer's undated edition of the *Legenda aurea*. Two rare French editions of

Boccaccio were bought late in 1890 or early in 1891: the 1483 Lyons edition of *La ruine des nobles hommes et femmes,* and Vérard's 1493 edition of *Les nobles et cleres dames.*

Morris's work in designing his new type continued throughout 1890, the last punches of the fount being finished off by Edward Prince in December. In July of that year Morris bought a copy of the 1527 Wynkyn de Worde edition of the Golden Legend, which gave its name to the type, and which Morris planned as his first production in it. In September he and Ellis entered into an agreement for the book with Bernard Quaritch, who would act as publisher. Morris was now fifty-six years old and from this time henceforward the Kelmscott Press and the library, as dual components of a passion for the well-designed, well-decorated book, would dominate his interest.

Many of Morris's purchases and prospective purchases of books were first aired to his friends in the S.P.A.B. The S.P.A.B. committee met regularly on Thursday afternoons in Buckingham Street, and when the sessions were over the men would cross the Strand to Gatti's restaurant where they ate and talked, the emphasis greatly on the latter. Sydney Cockerell, then in his early twenties, had known Morris slightly for several years through socialist meetings, but the acquaintanceship ripened when he was elected to the committee of the S.P.A.B. in March 1890. From then on his diary gives many interesting vignettes of Morris in action. On 14 August 1890 Cockerell recorded, 'W.M. at S.P.A.B. and Gatti's. 11 letters of the new type already cast. Went back home to Kelmscott House with W.M. and sat talking till 11.30. He had just bought a Sweynheim and Pannartz Suetonius with woodcut ornaments and initials. . . .' This beautiful Suetonius is now in the Morgan Library: the ornaments and initials Cockerell mentions are found in only a few copies of the book, for they were impressed by hand afterwards, and not as part of the printing run. The book is entered in the 1890–91 catalogue at just the point its date of acquisition would suggest. The following January, at Gatti's, 'W.M. had with him Veldener's Speculum humanae salvationis which he thought of buying.' He did not think for long: this rare illustrated Kuilenburg imprint is duly entered in Morris's catalogue, and several days later Morris wrote exuberantly to J. H. Middleton of the Fitzwilliam Museum telling of his new acquisition.

One great area of book art was still very sparsely represented in Morris's library: medieval manuscripts. Only seven early manuscripts are in the 1890–91 catalogue, and they make up a very mixed bag. Two Bible manuscripts, as we have said, seem to be holdovers from before 1876. There was a seventeenth-century Shah Nameh (a text in which Morris had some interest), later belonging to Wilfrid Scawen Blunt and now in the Fitzwilliam Museum; a humanist manuscript of Poggio, *De varietate fortunae*; and a fourteenth-century Paraldus, *Summa virtutum et viciorum*, in an early binding. The best manuscript in the group was of textual interest: an early fourteenth-century copy of the *Roman des sept sages* with two continuations, the *romans* of Marques and Laurin. This belonged formerly to Baron Seillière; Morris bought it from Quaritch in April 1890, and it is now in the Fitzwilliam Museum as part of the McClean bequest. One other manuscript was in an important binding, a gold-tooled Roman binding for Cardinal Giuliano della Rovere (later Pope Julius II), c. 1484, which is perhaps the earliest-known example of gold tooling from that city. William Morris acquired this at or shortly after the Crawford of Lakelands sale of 12 March 1891; recently it was in the collection of Major J. R. Abbey. Morris's collecting tastes never focused sharply on bindings as such, but as we shall see he owned several other pre-1500 bindings of outstanding quality.

The scarcity of early manuscripts in Morris's library to this date was not due to any lack of interest. To the contrary, Morris's fascination with medieval manuscripts was deep-seated, going back to Oxford days and the influence of Ruskin. His first known pieces of decorated book work, such as the 'Guendolen' leaf of 1856, were pastiches of medieval illumination. Mackail has quoted a tantalizing fragment from a letter of Morris's dating to October 1858, relating that he was in France again (he had been there in August with Charles Faulkner and Philip Webb) 'to buy old manuscripts and armor and ironwork and enamel.' What consequence this may have had we do not know. Morris and Burne-Jones were both frequent visitors to the Manuscript Room of the British Museum. A small notebook of Morris's from the Red House days has survived, filled with sketches from British Museum manuscripts mixed, in happy randomness, with caricatures, notes, decorations, and other rough drawings related to the work of Morris, Marshall, Faulkner & Co. But the progression from admiring objects to determining to possess them is a fateful one, which operates by its own peculiar psychology. We may surmise, on the basis of scattered evidence, that somewhere around 1891 Morris came to realize that his library was an important one, and that medieval manuscripts should naturally be ranged alongside its incunabula. From his own observations and from his talks with Emery Walker he was sensitive to the continuity from manuscript to early printing, both in letter and in decoration. He was aware that the striking woodcut ornamental initials of his favorite books, the early Ulm and Augsburg imprints, had models in manuscript illumination, sometimes in schools centuries earlier than the fifteenth century. His own work in the arts had always been attuned to the notion that design and decorative conventions evolve from tradition.

Manuscripts are unique objects, and so it is inevitable that, on the average, they will be more expensive than printed books of equal quality and importance. But in the 1890s great manuscripts and printed books alike were available in a deceptive profusion that tended to hold down their prices. In every year of the 1880s and 1890s the London book market saw a new flow (and, not infrequently, a recycling) of first-quality materials from private collections. By far the strongest single supporter of this abundant market was Bernard Quaritch. Though Morris bought from a number of dealers, a considerable part of his custom in the 1890s was given to Quaritch; the profits Quaritch owed Morris for sale of the Kelmscott Press books returned to the shop to pay for Morris's antiquarian purchases.

Perhaps the earliest indication of Morris's growing involvement with medieval manuscripts comes in April 1891, at the time of the second Edward Hailstone sale—shortly, that is, after Morris stopped making entries in the 1890–91 library catalogue. Messrs. Leighton bid on a number of lots for Morris, and, in a letter to them dated 30 April, he expressed his overall satisfaction with the outcome of the bidding. A small Psalter that went for £16 10s was, he said, 'cheap commercially, and of great value to me.' He was also pleased with another Psalter and a Book of Hours they bought for him. He was surprised, however, at the price fetched for a printed Book of Hours, 'all daubed up as it was. I now see that MSS. are much more uncertain at a sale than printed books'—which tends to confirm the impression that this was a new venture for him. He also asked that they reserve for his inspection a third Psalter. Besides these manuscripts got by commission to Leighton, at least four more Hailstone manuscripts and several early printed books came eventually into Morris's possession.

A group of six auction catalogues from the 1890s formerly belonging to William Morris has been preserved in the Grolier Club. His extensive annotations in the catalogues give a lively sense of his participation in the sales and of how he went through the books at the viewings. He would draw a bold line with a thick blue pencil down the pages containing nothing of interest. When he came to an item—usually a manuscript—of potential interest, he marked it with a vigorous X, XX, or XXX and with the word 'see.' Morris's comments on the manuscripts were rapidly formed and definite in tone (commas have been added for clarity): 'drawings very good, modest book with little color c. 1480'; 'very good all round book of date, c. 1350'; 'XX I want to see this, fine, big, red rubbed otherwise good, c. 1250, quite fragmentary'; 'XXX swell looking book very bright & clean—Italian at end otherwise a French book c. 1320. Very little picture work, very good crucifixion, writing fine'; 'lacks calendar, a very fine book, writing good, Bruges or Ghent: has St Bavon in Litany, c. 1260, but quality of illum. like English.' Against the printed Book of Hours in the Hailstone sale whose price he had wondered at to Leighton, Morris wrote 'sold £100 the painted pp. frightful daubs; the price absurd.' The general character of these notes accurately reflects Morris's taste in that he was not strictly or entirely a connoisseur of miniatures, but looked rather for a more subjective overall fitness of script, decoration, and illumination. His taste was formed not from scholarship, his own or others', but from decades of designing, often in a medieval idiom, in a wide variety of media.

Frequently Morris added marginal notes of the estimates, and these were sometimes, after the sale, corrected to the prices realized. Thus, in the Batemans' sale of 25 May 1893, Morris noted against an early thirteenth-century English Psalter 'X See £~~100~~ 115 too much.' That is, it was worth seeing and was estimated at £100, but went for £115, which was excessive. This might lead one to assume that Morris did not get the manuscript, as it went ('too much') for more than he was willing to pay. In fact he did acquire it: this was the so-called Nottingham Psalter, now in the Morgan Library and reassigned to Reading Abbey. One section of the Psalter, a *Laudes beatae Mariae virginis* found also in other manuscripts, was reprinted by the Kelmscott Press in 1896, as their first piece of three-color work.

Collectors often, perhaps usually, think that the prices they pay are too high; their protests may be taken to mean not that they intend to stop buying, but that if prices were lower they could buy even more. Morris's marginal complaints, 'too dear,' 'much too dear,' 'rather much,' 'too much,' 'very high,' appear throughout the catalogues, but he acquired as steadily as he complained. His attempts at bargaining were few and feeble. May Morris records that when Cockerell succeeded in reducing the price of one manuscript by £50 (or £25 in her version), Morris remarked ungratefully, 'Well, but you might have lost me the book, you know!' Shaw tells a similar story: Morris once paid £800 for a manuscript, and was admonished by someone in his company (Shaw does not identify this unsympathetic character) because it could have been bargained down to much less. When Morris 'was looking thoroughly miserable and ashamed of himself for his weakness, I [Shaw] said, "If you want a thing you cannot bargain." He instantly recovered his selfrespect; and his gratitude to me was boundless.'

One flaw in Shaw's story is the assumption that this particular manuscript could have been bought for less. This was the late twelfth-century Huntingfield Psalter, in the hands of Bernard Quaritch, which it took Morris almost three years to decide to buy. He first had it on approval in August 1892, when he showed it to Cockerell and Walker. But at this time Morris had not yet spent anything in the neighborhood of £800 for a book or manuscript, and he could not bring himself to take the plunge. The Psalter appeared more than a year later in Quaritch's *Catalogue of manuscripts arranged in chronological order* (December 1893), alongside many other splendid items. Morris did not make a move for the Psalter, though he did eventually buy at least three other manuscripts in this catalogue. But he could not forget the Huntingfield Psalter, and in July 1894 he had it on approval from Quaritch again. At the end of that month he wrote to Cockerell, 'Kindly take the Huntingfield to Q in a day or two. I have told him that I am not going to buy it.' Several weeks before, at the Fountaine sale, Morris had just bought the Grey-Fitzpayn Hours for some £425, which seems temporarily to have cooled his fever. But Quaritch may have realized that he was making progress, for nine months later he dangled the Psalter before Morris once more. On 29 April 1895 Cockerell responded that

> Mr Morris, who has just returned from the country, asks me to thank you for offering him the refusal of the Huntingfield Psalter, & to express his regret that he does not feel able to become the purchaser of the book.

This sounds quite definite, but was not. Three days later Cockerell's diary records, 'W.M. went to Quaritch's with F. S. Ellis and bought the Huntingfield Psalter. He was at S.P.A.B. and Gatti's.' Morris wrote his initials and the date, May 2nd 1895, on a flyleaf of the volume, perhaps to insure that he could not return it again. Increase of appetite grows by what it feeds on, and little more than a week later Morris did not hesitate in acquiring, for £900, the beautiful Tiptoft Missal, which stands with the Huntingfield Psalter in the small group of his finest manuscripts.

In October 1892, during a memorable visit to Kelmscott Manor, Sydney Cockerell was asked by Morris to help in cataloguing his library; the stipend was later worked out at two guineas a week and a copy of the Kelmscott *Golden Legend*. Morris estimated, in his optimistic way, that the work 'would take three or four months at least.' At the time the offer was made, Cockerell was a young man of twenty-five, still not settled in a career. Since he was seventeen he had worked as a clerk in the family firm of coal merchants, a position that could not have been more greatly at odds with his natural inclinations. In the 1880s Cockerell had cultivated an acquaintance with Ruskin (originating from a gift of shells to the master) into a loyal friendship, and had discovered in himself a passionate interest in architecture. Following a legacy from his grandfather in 1889 he had considered going up to Cambridge, but was dissuaded by family friends. At the close of 1891 he sold his interest in the firm, in which he had invested his legacy; after dividing the proceeds equally with his brothers and sisters, and paying out of pocket a family debt of honor to one of the firm's clerks, he was left with less than £100. The final resignation from the coal business came in May 1892, and for the succeeding months, until Morris's welcome offer came along, Cockerell had no fixed occupation.

Cockerell had not previously shown any special interest in early printing and manuscripts, except insofar as he shared in the frequent viewings of Morris's books on Thursday evenings at Gatti's and at Kelmscott House. To some extent the offer must have been made as a favor, to help out at a time of financial uncertainty. Cockerell himself gave much of the thanks to Philip Webb's influence. The day before he began as Morris's librarian, as if to inspire himself for the task ahead, Cockerell spent several hours at the British Museum, studying the exhibit of incunabula in the King's Library. The next morning, at Kelmscott House, F. S. Ellis was there to give Cockerell hints in cataloguing. Ellis had retired from bookselling in 1885, putting up his stock at Sotheby's and passing on the goodwill of the firm to his nephew, Gilbert Ellis. His association with Morris remained close, and he was a frequent visitor to Kelmscott House.

The new job was congenial to Cockerell: 'never was so happy in my work before,' he recorded three weeks after starting. Not least of its attractions, naturally, was the proximity to Morris. Cockerell's admiration for him bordered on reverence, and only grew stronger with the years. Morris's love of books was shared especially with a small group of friends who were the most frequent visitors to the library: Emery Walker, F. S. Ellis, and Charles Fairfax Murray when he was in England. Cockerell now joined this inner circle of bibliophiles. Philip Webb and Edward Burne-Jones also came by to see particularly fine manuscripts, but they were less interested in early printing and in the news and gossip of the rare book trade. One day, Cockerell recorded, Ellis was in the library in the morning, and Fairfax Murray in the afternoon, 'and the talk, though very interesting and delightful in itself, did not assist me in my work of cataloguing.'

In addition to the cataloguing proper, there was much organizing to be done, as the confusion of the 1890–91 catalogue corroborates. Shortly after he began, Cockerell spent two full days in bringing order out of chaos, with 'a great turning out of drawers etc. and general rearrangement—et quorum pars magna fui.' Further rearranging was done in 1893, when a new bookcase was bought for the drawing room, to display the handsomest bindings. As books continued to pour in, more shelves were put up in Morris's bedroom. The cataloguing itself often went very slowly because of the complications of the woodcut books. Cockerell made a census of cuts in each volume, with notes of the number of repeated cuts. The Nuremberg Chronicle, with its hundreds of woodcuts and hundreds of repetitions, was a special problem. In January 1893 Cockerell 'spent nearly the whole day' on it, 'and finally gave it up in despair of ever counting and sorting the cuts.' He left the book alone for over a year, until on 7 February 1894 his diary records, 'Attacked the Nuremberg Chronicle.' On 12 March he 'resumed work' on it, on the 27th he 'at last [began] to see the end,' and the next day he finished. The library never was completely catalogued. For the last nine months of 1893 Cockerell served as secretary at the New Gallery, doing only occasional work at Kelmscott House on weekends. But aside from that sabbatical he remained in Morris's full employ, adding to his duties in July 1894 that of secretary of the Kelmscott Press.

It may be, though positive evidence is lacking, that even before he hired Cockerell, Morris was gestating the plan of issuing a printed catalogue of his early books and manuscripts, as a Kelmscott Press publication. The 1890–91 catalogue was very inadequate and steadily becoming more so, even as a rough overview of the collection. Concurrently with that catalogue Morris had begun but never finished a more formal catalogue of his important books. This he wrote out in a calligraphic script, as one of his last endeavors in calligraphy. A gathering of eight leaves is all that was done, describing a few early manuscripts, several of his own calligraphic manuscripts, and fifteen of the earliest incunabula, mostly Italian, none illustrated. This calligraphic catalogue contains colophon transcriptions and notes of decoration in the copies and of general condition; there are no indications of format or collation.

Cockerell's work was of course more elaborate. A considerable amount of his cataloguing at Kelmscott House has been preserved. The descriptions of the illustrated incunabula are often found laid in or pasted in the books themselves. There is a summary catalogue in his hand of 'Early books with woodcuts' which he entered, arranged alphabetically by printer, in a ledger book. With few exceptions, these are books printed before 1520. Cockerell also made a similar catalogue, on loose leaves of ledger paper, of the unillustrated early printing; and finally there have survived a number of his rough notes, some annotated also by Morris. Cockerell's work at Kelmscott House led him into a wider acquaintance in the book world, and the library-related entries in his diary, which at the beginning are sometimes rather naive, show a steadily growing expertise. For the first time he became a reader at the British Museum, and he began attending meetings of the recently founded Bibliographical Society. At various times he consulted with W. H. J. Weale, the learned liturgiologist of the South Kensington Museum, and with Robert Proctor of the British Museum, whose researches in incunabula and early typography were epochal. Proctor was a year younger than Cockerell, but already a finished scholar with an extraordinary capacity for work. He visited Kelmscott House several times for 'tea and books,' and many of Morris's enthusiasms became his own: S.P.A.B., the Kelmscott Press, modern type design, and Icelandic literature. Some of the attributions of the 'sine nota' books in the library were made by Proctor, and Morris also acknowledged his help in his article on Augsburg and Ulm woodcut books.

In May 1894 Cockerell recorded that Morris had begun his notes for the printed library catalogue, which was listed among the forthcoming books in the Kelmscott Press announcements. A good many of these notes are still extant, laid into the books they relate to; others have strayed from their volumes. Had they ever been completed, they would have made up a kind of autobiography of Morris as a book lover, for his reactions to the books at hand come through with great clarity. They depend, often, on Cockerell's more detailed studies, but their concentration is almost entirely on elements of design. The notes concerning incunabula are numerous, but only four notes by Morris on his manuscripts are known, all concerning Psalters now in the Pierpont Morgan Library. The work on this catalogue was nowhere near completion at Morris's death, but he had begun planning the illustrations, and Emery Walker made some dozens of electrotypes of fifteenth-century woodcuts for this purpose. After Morris died Cockerell, with Proctor's assistance, brought these reproductions together in one of the Kelmscott Press's last publications, *Some German Woodcuts of the Fifteenth Century*. The preface was extracted from Morris's article in *Bibliographica* on the Augsburg and Ulm woodcuts, and there was a list, taken from Cockerell's cataloguing, of 131 important woodcut-illustrated incunabula in the collection. As if to commemorate his own most exhaustive piece of work for the library, Cockerell included his 'Analysis of the Woodcuts in the Latin Edition of the Nuremberg Chronicle.'

Morris prepared several lectures and essays in the 1890s that were based more or less directly on his own collections. The most specialized of these was his article 'On the Artistic Qualities of the Woodcut Books of Ulm and Augsburg in the Fifteenth Century,' commissioned in 1894 by the editor of *Bibliographica*, A. W. Pollard. This essay, which Cockerell tells us was delivered up 'not without murmuring,' was in a sense the most personal of Morris's writings on the book arts in that it is founded almost entirely on the direct examination of his favorite group of printed books. His summary of the aesthetic of the Ulm and Augsburg woodcuts is revealing:

> Their two main merits are first their decorative and next their story-telling quality; and it seems to me that these two qualities include what is necessary and essential in book-pictures. To be sure the principal aim of these unknown German artists was to give the essence of the story at any cost, and it may be thought that the decorative qualities of their designs were accidental. . . . I do not altogether dispute that view; but then the accident is that of the skilful workman whose skill is largely the result of tradition; it has thereby become a habit of the hand to him to work in a decorative manner.

Morris had already made this point in a slide-illustrated lecture given at several times on 'The Woodcuts of Gothic Books': the cuts in fifteenth-century books performed a dual function, of illustrating their text simply and directly, and of decorating the pages they appeared on. To harmonize these two functions involved a difficulty, but 'it was just that kind of difficulty or limitation which made an art worth following.' These words are interesting in that Morris could equally well have been writing about the Kelmscott Press and about the effect he was striving for there. This combination of ornament and structure was requisite for Morris in all the arts he practiced or studied; he wrote elsewhere of 'the cant of the beauty of simplicity.' It was always Morris's special genius to communicate a sense of unity through his many seemingly diverse enterprises. Nowhere is this continuity more strongly felt than in the relation between his library and the Kelmscott Press.

In March 1895, at the time of the seventh Phillipps sale, Morris reported to F. S. Ellis on the manuscripts he had lost to higher bidders, and wrote, 'rejoice with me that I have got 82 MSS., as clearly I shall never get another.' He was soon proven wrong. Little more than a month later, it will be recalled, he acquired the Huntingfield Psalter and the Tiptoft Missal in rapid succession. And despite Morris's mock prediction, still more irresistible material continued to come his way until eventually his collection totaled 112 manuscripts. In the summer of 1895 he bought a fine French thirteenth-century Book of Hours from Jacques Rosenthal in Munich for £450. This was the manuscript whose price Cockerell reduced by £50, without earning Morris's gratitude. Morris also got at this time a very attractive fourteenth-century *Roman de la Rose* with numerous grisaille miniatures (£400, Quaritch). In the autumn he paid £325 to Ellis & Elvey for a Psalter, £650 to Leighton for a large folio French thirteenth-century Bible in three volumes, and £200 for a large folio French Josephus, formerly in the Hamilton Palace collection. Just before Christmas he rounded out the year with another Psalter from Ellis & Elvey for £375 (Cockerell believed it was overpriced).

For over a year before his death Morris suffered from a progressive weakness, different from his usual attacks of gout. It became more pronounced after the New Year, 1896, when he began losing weight steadily. A succession of examinations by several doctors did not produce any conclusive diagnosis until the summer, when they finally affirmed that it was tuberculosis. Morris's interest in medieval manuscripts was stronger than ever during these last months, but the history of his collecting now carries an undertone of poignancy, though it involves several of his greatest acquisitions. A set of entries in Cockerell's diary tells the story of one purchase with great directness: on 28 February, 'W. A. S. Benson and Fairfax Murray to tea' (Benson was an architect and designer whom Morris had known for many years); on 2 March, 'F. S. Ellis, Fairfax Murray & W. A. S. Benson (with his fine 12th century N. Testament) at K.H. to tea'; and on 9 March, 'W.M. going on well—He has bought the Benson New Testament for £250.'

Toward the end of April Morris received in the mail from Jacques Rosenthal an enticing photograph of an English Bestiary that Rosenthal considered too precious to send on approval. Morris replied the same day that Cockerell might go to Munich to see it, with power to buy if all was well. Fairfax Murray was then about to leave for Venice, and he too planned to stop first in Munich to see the Bestiary. After Murray left, Rosenthal wrote that the manuscript was now in Stuttgart, and Cockerell left for there the next morning, 28 April. He reached Stuttgart the evening of the 29th; the following day he saw Rosenthal, bargained over the Bestiary, handed Rosenthal a check for £900, and left the same afternoon with the manuscript in hand. He arrived in England on 3 May, and the next morning showed the manuscript to Emery Walker before taking it on to Kelmscott Manor. His friend Jane Duncan came with him as far as Oxford, where Cockerell paused to compare the manuscript with two fine Bestiaries in the Bodleian (one of which must have been Ashmole MS.1511). He reached Kelmscott at 7.30, where F. S. Ellis was also staying, '& the book voted by everyone a magnificent one—W.M. greatly pleased.' Morris and Cockerell returned to London the next day, and Morris never saw Kelmscott again.

In June and July Morris made another spectacular acquisition through a curious set of circumstances. He had been a member of the Society of Antiquaries since 1894, having been elected according to Cockerell (who must have paraphrased Morris's own explanation) 'in order that he might be part owner of the Lindsay Psalter'—the early thirteenth-century Psalter made for Robert de Lindesey, Abbot of Peterborough, one of the Society's greatest treasures. In June 1896 the Society mounted an exhibit of 'English mediaeval paintings and illuminated manuscripts,' to which Morris was asked to contribute. At the beginning of the month he and Cockerell took six of his manuscripts, plus a four-leaf fragment, to Burlington House for the exhibit. The fragment, splendid in its own right, was from a late thirteenth-century Psalter, otherwise seemingly unknown. The leaves had belonged to Fairfax Murray, but in 1894 Morris persuaded him to make a trade for a fragment of eight leaves from a fifteenth-century Italian picture chronicle, recently acquired from Quaritch.

Two days after delivering his manuscripts, Morris returned to Burlington House with Burne-Jones to see the exhibit, which was filled with outstanding items. Morris found his four leaves exhibited beside a magnificent Psalter that he immediately recognized as the fragment's source. This Psalter belonged to Henry Hucks-Gibbs, first Baron Aldenham. Not only was it magnificent, and the source of his fragment, but Morris also recognized it as the same Psalter he had examined almost twenty years before, when Aldenham had left it in F. S. Ellis's shop. Morris had then pronounced it to be 'one of the most valuable books of English art in existence.' Now he was determined to possess it; as he wrote to Ellis the next day, 'the question is, How am I going to get hold of the Hucks-Gibbs "fragment"?'

In the meantime the next installment of the Phillipps manuscripts came on the block, and Morris made two acquisitions at bargain prices. One was a fine twelfth-century St. Albans Psalter, knocked down for only £34. The other, even more remarkable, was a Hegesippus in a blind-stamped Winchester binding of c. 1150; severely underdescribed in the catalogue, it went to Morris for only £13 (he had asked Quaritch to bid to £150 on it). Less than a handful of Winchester bindings, the earliest decorated English bindings, survive, and of these the Phillipps example is the finest.

Lord Aldenham's 'fragment' was not the only manuscript at Burlington House that Morris coveted. He fell in love also with the Psalter belonging to the Duke of Rutland, one of the great achievements of English Gothic illumination; here too, he wrote Ellis, 'I am beating my brains to see if I can find any thread of an intrigue to begin upon, so as to creep and crawl toward the possession of it.' Two separate sets of negotiation now began. St. John Hope, assistant secretary of the Society of Antiquaries, approached Aldenham, suggesting as a hint that one of the two men should sell to the other. This line did not produce fruitful results, as by reason Morris should then surrender his four leaves to Aldenham for some modest sum. Finally, on 4 July, Morris wrote Aldenham directly, offering him a flat £1000. This offer succeeded, and two days later the manuscript was in Morris's hands, Aldenham recording in his diary, 'seeing that it had cost me £75, 36 years ago, I could not but accept.' Because of the windmill depicted on the second folio, Morris named this, as it has since been known, the Windmill Psalter. It was his last recorded acquisition.

Approaches to the Duke of Rutland concerning his Psalter were made by R. H. Benson, son-in-law of the great collector Robert Stayner Holford. Discussion went on for several months, Morris offering an even greater sum than for the Windmill Psalter, but in mid-August the Duke finally decided he could not bear to part with it. In late July, on his doctor's advice, Morris was sent on a cruise to Norway, in the vain hope the northern air would help his lungs. It was not a happy trip; Morris felt tired and melancholy, and he missed his library. When he returned a month later he was weaker than when he left.

He knew now that he would not recover, and the next five weeks at Kelmscott House were a waiting for death. Even then, though he could no longer walk and occasionally hallucinated, Morris could not be inactive. He finished his last book, *The Sundering Flood*, by dictation, and continued to design ornaments for the Kelmscott Press. But his manuscripts were his final pleasure; he looked at them when he was too weak for any other activity. R. H. Benson brought by three of the Holford manuscripts from Dorchester House, all old favorites of Morris's. These included the Bible Historiée with almost a thousand miniatures that he and Burne-Jones had looked through together; and the Psalter made for Yolande Vicomtesse de Soissons, whose connection with Amiens Morris had been the first to notice. The Duke of Rutland also sent his wonderful Psalter for Morris to see for the last time. It was Sydney Cockerell's recollection that the St. Albans Psalter from the Phillipps sale, which Morris called the Golden Psalter, was the last manuscript he looked at and spoke about. On the morning of 3 October, William Morris peacefully died.

F. S. Ellis and Sydney Cockerell were co-executors, with Mrs. Morris, of the estate. A major part of their work involved the library, whose valuation for probate was begun by Ellis and his son four days after the funeral. One of the best manuscripts, the Grey-Fitzpayn Hours, was taken up to Cambridge. After Morris had acquired it, Emery Walker recognized one of its missing leaves in the reproductions he was printing for M. R. James's catalogue of the Fitzwilliam manuscripts. As was found, two of the leaves were there, bought by Samuel Sandars at the 1892 Lawrence sale. A generous agreement was worked out, whereby the leaves were restored and the volume sold to the Fitzwilliam for £200, Morris keeping possession of it in his lifetime.

Fairfax Murray expressed a desire both to take over Kelmscott House, and to acquire Morris's library *en bloc*. In December Cockerell and Ellis made a more detailed appraisal of the library. They computed the value of the manuscripts at about £12,500, and of the early printed books, which now numbered some 800, at £8000. A tentative agreement was formed to sell the entire library to Murray for £20,000. He paid £2000 immediately and took away some two dozen books and manuscripts, including the Windmill Psalter, with a proposal to pay the balance in installments of at least £3000 a year. On reflection, however, Ellis found it necessary to rescind the agreement, on the grounds that Murray could not put up proper security for the unpaid balance. This was done amicably, with no impairment to the friendship of any of the three men.

In February 1897 E. Gordon Duff made inquiries as to purchasing Morris's books on behalf of the newly founded John Rylands Library. A problem here may have been that while the manuscripts would have made a very significant addition, the printed books would be in great part duplicated by the Spencer collection there. Other visitors during these weeks who may have had an interest in purchasing the library include L. W. Hodson, Henry Yates Thompson, Lord Balcarres, and Dr. Friedrich Lippmann, of Berlin. Representatives from Messrs. Pickering and Chatto and from Sotheran also called to inspect the books. Bernard Quaritch was involved in the attempt to keep the library intact. In March 1897 he printed an unpriced and rather reticent catalogue of *The Best Books in the Library of the Late William Morris*. This was a selection of 75 manuscripts and 249 printed books to which Quaritch added a note stating that 'Should the Library not be sold *en bloc*, I shall be glad to purchase any of the above for my clients.'

The most serious signs of interest came from Messrs. Pickering and Chatto, who made several visits. One day in early April Chatto brought a client to look at the books, but apparently did not introduce him to Cockerell. Two weeks later a firm sale to Pickering and Chatto was concluded, for £18,000. The last days of April were spent in packing up the books, 'a dismal business,' Cockerell noted in his diary. The sale included all of Morris's Icelandic books and several hundred other later volumes that Mrs. Morris did not want. Pickering and Chatto did not divulge the name of their client. Only three months later did Ellis learn his identity, from Bernard Quaritch: a Mr. Richard Bennett, of Pendleton near Manchester. Ellis himself had, he wrote, been 'positively told that Mrs. Rylands was the purchaser & on another occasion with equal certainty of knowledge that the Empress Frederic of Germany had bought the books.' In any case, he wrote Quaritch, 'I have every reason to believe that the purchase was made quite independently of your excellent list.' This is almost certainly true, as Chatto's first visit to Kelmscott House was made before the list was issued.

Richard Bennett, a Lancashire manufacturer, was then and still remains a very obscure figure. At the time he bought Morris's books he was already building a splendid library, including many Caxtons. All his buying seems to have been done in the 1890s, with Pickering and Chatto as his agents in the great sales. Seymour De Ricci qualified Bennett as eccentric, and he and Cockerell both affirmed that Bennett had a special dislike of large folios, De Ricci setting his limit at thirteen inches. Bennett did in fact possess a few volumes larger than this, but only a very few. He had a highly selective attitude toward his library; from Morris's collection he kept only 31 manuscripts and 239 printed books. The remainder was auctioned at Sotheby's on 5 to 10 December 1898, in 1215 lots; the sale realized just under £11,000. The importance of many of the illustrated books was very imperfectly conveyed in the sale, 'spirited' being almost the only adjective used by the cataloguer to describe the woodcuts.

Many, perhaps even most, of the lots in the 1898 sale were not of great intrinsic interest, but some were of such importance that it is hard to conceive why any bibliophile should have wanted to dispose of them. Of the thirty-five manuscripts appraised by Cockerell and Ellis at £100 or more, Bennett kept twenty-one. He was very stern on books printed after 1500, so that he retained only a handful of Morris's important early sixteenth-century imprints. A number of outstanding large folio items were sold, including the Sherbrooke Missal, now in the National Library of Wales. A very fine 1478 binding by the binder and illuminator Ulrich Schreier, made for the Archbishop of Salzburg, was disposed of for no apparent reason except that it was sixteen inches tall. Another volume inexplicably put up for sale was the Winchester binding. When Ellis saw it in Sotheby's catalogue he wrote Cockerell, 'what a dolt and idiot must that Bennett be to know no better than turn out that Hegesippus.' This great binding was acquired by Henry Yates Thompson, passed into the collection of C. W. Dyson Perrins, and after the Second World War was very appropriately donated by Perrins to Winchester Cathedral.

Richard Bennett did not hold onto his library for long. In 1900 he issued, in only a few copies, a brief little printed Catalogue which must have been the prelude to an attempt to sell. It listed 559 printed books and 107 manuscripts, Morris's library contributing just over forty percent of the total. In the summer of 1902 Pierpont Morgan purchased Bennett's books. Almost nothing is known of the circumstances of the sale, though it has been said that the price paid was £140,000. Following the purchase, Morgan commissioned a sumptuous, heavily illustrated catalogue in four volumes of his early printed books and manuscripts, not all of which came from Bennett. The manuscripts were described by M. R. James, and the printed books by E. Gordon Duff, Stephen Aldrich, Robert Proctor, and A. W. Pollard. Proctor especially was interested in the assignment because of its connection with Morris; and so it has happened that this elaborate project of an American archcapitalist remains the best single guide available to William Morris's love of old books.

The story of Morris's library is a very attractive one. His was not the finest collection of its time, not even, probably, the finest collection then in London. According to several measures Morris fell short of bibliophilic standards: he did not look on imperfect copies with the revulsion they deserve, he cared nothing for rarity, and there was no book he felt compelled to acquire except according to his own tastes. The concept of a 'complete' collection was alien to him. When he wanted, for instance, a particular woodcut or border decoration, Morris could be as happy with a later occurrence as with the first. Almost all his knowledge of books and manuscripts came from a very sharp eye and a powerful memory, rather than from research. But few men can have gotten greater or purer pleasure from their libraries. Morris's book-buying was, as his daughter wrote, 'a great passion, not exactly the collector's passion.' His love was grounded in his feeling of kinship with the long-dead generations of craftsmen who created these beautiful objects. Morris was quite untouched by the commonest blemish of collectors, the belief that one gains in personal importance according to the value of the objects one owns.

His books were at the service of his friends, and his greatest pleasure was in sharing new finds. The careers of Emery Walker and Sydney Cockerell took new, decisive, and highly influential turns through their happy association with Morris and his library. Charles Fairfax Murray, one of the great collectors of modern times, dedicated his catalogue of early German books to the memory of Morris, 'to whose intimate knowledge of early books I owe my first inspiration as a collector.' When William Morris's library is spoken of, the words ought always to be silently added, 'et amicorum.'

William Morris: Calligrapher

TO WILLIAM MORRIS THE IMPORTANCE OF BEAUTY WAS so vital that he devoted his life, his energies, and his considerable talents to the art that creates and sustains it. The desire to produce beautiful things was, he tells us, a leading passion of his life. When he spoke and wrote of the 'pleasure of the eyes' and of the art that brings joy to the maker and to the user of an object, he was speaking out of his own extensive experience as a master craftsman and designer who urgently wished to raise public taste so that it would no longer tolerate shoddy or ugly work in any product, large or small.

His enthusiasm for good design and good workmanship caused him to admire and study the craftsmanship of various countries and different periods of history, especially that of the European Middle Ages. Medieval architecture and illuminated manuscripts held especial meaning for him:

> I remember as a boy going into Canterbury cathedral and thinking that the gates of heaven had been opened to me, also when I first saw an illuminated manuscript. These first pleasures which I discovered for myself were stronger than anything else I have had in life.

In manuscripts he caught glimpses of medieval life; they abounded in ornament that often made use of natural forms (another of his enthusiasms); and they provided evidence of the skill of the scribe and the lively imagination of the artist. All of it was presented in vivid color (in contrast to the starkness of print) by a workman who 'lived amidst beautiful works of handicraft and a nature unspoiled by the sordidness of commercialism.'

While he was a student at Oxford, Morris enjoyed studying the medieval manuscripts at the Bodleian Library, and after settling in London he visited the British Museum for the same purpose. In 1856, the first year that Morris and Burne-Jones roomed together in London, the former tried his hand at writing and decorating manuscript pages in which the medieval inspiration was obvious. His attempts at this time were part of his endeavor to discover the direction in which his artistic talents lay; a few years later he found it as a prolific pattern designer for, and chief executive of, Morris, Marshall, Faulkner & Co. His new friend Dante Gabriel Rossetti wrote of these early efforts: 'In all of illumination and work of that kind he is quite unrivalled by anything modern that I know—Ruskin says, better than anything ancient.' This is a reference to the great popularity, especially in the 1850s and 1860s, of do-it-yourself illumination. In the 1840s men like Owen Jones and Henry Noel Humphreys published books describing and illustrating medieval manuscript lettering and decoration, and designed volumes in which a text in gothic lettering was surrounded by colorful chromolithographed borders of medieval inspiration. The number of manuals for home use was at its peak in the sixties. Charlotte M. Yonge, for instance, in her novel *The Trial* (1864) refers to a young lady who had 'an illuminator's guide' at home and 'a great deal of red, blue and gold paint.' This activity was stimulated by the Gothic Revival and the renewed interest in religious art. Consequently the texts chosen tended to be Biblical or otherwise inspirational, and the lettering commonly was gothic.

Though Morris used a form of gothic script for the three examples he is known to have done at this time, his texts were drawn from quite different sources. The occasion of one of them, perhaps the first, was the visit of Robert and Elizabeth Browning to England in the summer of 1856. Morris and Burne-Jones met them through Rossetti, and Morris was inspired to present them with two stanzas of a song from *Paracelsus* (lines 190–205, version of 1849) in medieval form. His script is a rather weak gothic letter of vertical strokes with 'feet' on them. The dark, crowded lettering of the stanzas is set off by large golden squares which form backgrounds for red and blue initials that owe something, but not a great deal, to Lombardic letters. Above, below, and between the stanzas slender slanting lines of red and blue, with short twigs and buds sprouting from them, provide a lightly decorated area on which the heavier verses seem to float. This area is bounded closely on either side by solid columnar borders which maintain the color scheme. From the top of each column protrudes a grotesque head displaying long flying ears in the manner of late thirteenth- or fourteenth-century manuscript decoration. The lower end of one terminates in a small head with flowing locks, and the other in a series of leaves.

In the same gothic letter, but with more elaborately planned decoration, is his own poem 'Guendolen' which he illuminated for Georgiana Macdonald, later Mrs. Burne-

Jones, in August 1856. 'Guendolen' had been published under the title 'Hands' in *The Oxford and Cambridge Magazine* for July of that year and was shortly thereafter included in his longer poem 'Rapunzel' which Rossetti admired. Its six three-line stanzas begin with a dominating initial of red and blue accompanied by leafy spirals, one of which ends in a serpent's head. Each line of verse was to start with a colored initial on a colored background, but few of them were finished. Disembodied heads appear here and there, and below the last stanza a female figure with long hair holds a tress of it out over blue water. The border in this manuscript is a frame in which gold is used as well as red and purple-blue. Its elements are quite irregular. Leaves of varying size sprout from it, and here and there gold balls are associated with them. The decoration of this manuscript has been described as deriving from English fourteenth-century illumination which has absorbed Italian motifs such as the golden balls and the leaf forms.

The first part of 'The Iron Man,' a translation of Grimm's 'Der Eisenhans,' contributed the text for the third manuscript. This he gave to Louisa Macdonald, later Mrs. Baldwin, about 1857. The script is the same as that used in his other manuscripts, but this time, since he was writing prose, there is much more of it. He arranged it in two dense columns broken here and there by initials or by spaces for initials. Here again the text commences with an oversized initial, but this time the large Lombardic N contains a small painting of an incident in the story. The initial sends forth into the upper and left margins a series of tangled spirals, but there is no indication of a border. Instead, Morris decorated the space between the columns of writing with vines, leaves, and disembodied faces. This decoration ceases about halfway down the page where a large square blank space was left, possibly for a miniature. On this page also there are elements in the decoration that suggest Morris's interest in the decorative art of the fourteenth century. His biographer, J. W. Mackail, writing many years later, remarked that

> the design and colouring of the border and the treatment of a picture in the large initial letter, show a complete grasp of the principles and methods of the art. It is probable that no illumination had been done since the fifteenth century which was so full of the medieval spirit.

Morris's daughter May describes the foliage and grotesques as being 'in the early fourteenth century manner,' but she likens the heads with streaming hair to 'the unexplained dreamlike fragments of poems in the Guenevere time.' It is well to remember that while he was decorating these, and possibly other, manuscripts, the energetic young man was writing the colorful verses that he published in 1858 as *The Defence of Guenevere*.

From these early efforts in the book arts all the way to his Kelmscott Press pages, Morris's borders were rarely static or architecturally symmetrical. Growth, more or less controlled, was their chief characteristic. Though the growths in his later borders spring from the soil, these that he painted before his career as a designer began seem to combine the animal and the vegetable in forms unknown to biology.

In 1857 Morris joined Rossetti, Burne-Jones, and other artists in the joyful and short-lived venture of painting Arthurian scenes around the upper windows of the Oxford Union Society's debating hall, and in 1858 he painted Jane Burden, shortly to become Jane Morris, as a medieval maiden with a handsome manuscript on her bedroom table. The work of medieval artists was not his only stimulus, but it was particularly strong at this time and, in one form or another, it remained with him through his life.

Before Morris decided, a dozen years later, that the practice of calligraphy and manuscript illuminating was the best way to spend his time on Sundays, he and Burne-Jones had attempted to emulate the great printers of the fifteenth century by planning a large folio edition of his narrative poems that made up *The Earthly Paradise*. It was to have many wood engraved illustrations designed by Burne-Jones and cut either by Morris or by one of his friends. About forty-six blocks were actually cut for 'The Story of Cupid and Psyche,' but Morris was disappointed with the way they looked when trial pages were set up at the Chiswick Press in Caslon and in Basel type. Visitors to this exhibition may see the contrast between the weight of these forceful illustrations, so different in intention from the work of contemporary illustrators, and the type in which the text is set. As eventually published (1868–70), *The Earthly Paradise* had no illustrations other than a vignette on the title page: medieval maidens bearing medieval musical instruments, standing by a garden wall. Otherwise, typographically speaking, it resembled any other book of the period. A second try in the early seventies had the same results. For *Love Is Enough*, a long masque in verse, Morris designed several borders and two initials which he cut on wood, but again the project was unsuccessful.

In both of these attempts, Morris and Burne-Jones found that the illustrations and decorations that suited their taste overwhelmed the type faces available at the time.

Text type, for the most part, was thin and pallid, and Morris did not know how to surmount this basic obstacle. Furthermore, book publication was not the major concern for him then that it became twenty years later when his resolve to design a fount of type preceded his decision to found a press. At the Kelmscott Press, Morris controlled all aspects of design and production, but at this point he could only do so in manuscripts where he provided the letterforms as well as the decoration. In this, as in a number of other ways, his manuscripts are the direct ancestors of his 'typographical adventure.'

Thoughts of this kind may have been in Morris's mind in 1867 and 1868 when hopes for an illustrated *Earthly Paradise* were flickering out, for we find evidence that he was beginning to show an interest in forming letters as distinct from writing them in his ordinary longhand. An early example occurs in the handwritten pages of the Prologue (c. 1867) to *The Earthly Paradise* where five lines of the narrative suddenly show different characteristics from those that precede and follow them. The letters are more upright and less frequently linked within the words, while single strokes replace many of the loops in ascenders and descenders. An even more extensive incursion occurs in the midst of Morris's longhand translation of the *Eyrbyggja Saga* of 1869 (Fitzwilliam Museum) where this more studied script covers three full pages and parts of two others. They show that Morris had in mind the formation of clear, upright, separately written, sans serif letters, and was considering their massed effect on the page. It may have been Morris's first attempt to write a text in roman script. Some features, notably the ascenders that bend to the right, reappear in his calligraphy. Evidence of his interest in letters is again visible on the margins of his longhand manuscript of 'Bellerophon in Lycia' (1869), one of the narratives in *The Earthly Paradise*, which he was writing about the same time that he was working on his translation of the saga. As Morris wrote or revised his poetry, he frequently sketched flowers and foliage in the right-hand margins of his paper. His rare use of deliberately formed letters and words instead of blossoms and leaves indicates their importance to him at this time. On pages 50 and 51 they resemble the script used in the saga, and in some instances they look forward to his first calligraphic manuscript, *The Story of the Dwellers at Eyr*—another treatment of the same saga. In the later pages of 'Bellerophon' one can see the arrival of a script Morris used in several manuscripts: one which features thin diagonal strokes attached to the right legs of the letters h, k, m, and n. Not only does 'Bellerophon' illustrate the development of Morris's lettering in 1869, but it also suggests that he had become acquainted with the writing books of Ludovico Vincentino Arrighi (1522 and 1523) and of Giovanniantonio Tagliente (1525) which he may well have owned by this time. His treatment of ascenders may be derived from chancery script, and also the forms of several capitals, notably an N (page 83) with a descender from the right vertical stroke that curves beneath the letter like a J. The capital G has a similar scoop descender. On a long sheet containing part of 'The Story of Rhodope' (c. 1868), also from *The Earthly Paradise*, another distinct anticipation of his subsequent practice is clearly shown. Here Morris drew eight capital letters set in squares of single or double lines. Of most interest are three which were drawn with more care than the others. Not only do these letters (two N's and a D) have good proportions, but each is given a leafy background. They point directly both to the decorated initials he would soon be painting, and to his wish to bring the decorative qualities of growing things to his calligraphic pages.

In the autumn of 1868 Morris added to the demands of his business and his literary pursuits the study of the Icelandic language and its literature with Eiríkr Magnússon. The *Eyrbyggja Saga* was the first one they read together. Within a few months, Morris had gone through the bulk of the heroic literature, and had conceived a deep and permanent love for these northern tales, the persons that peopled them, and the spirit they represented. He visited Iceland in 1871 and 1873, published several volumes of translations, and found enjoyment in using a number of their texts for his calligraphy.

We know very little about how Morris prepared himself to write the scripts that one sees on his pages. His close examination of medieval manuscripts and his awareness of the sixteenth-century writing masters have been mentioned. Mackail reports that when he took up the art of illumination 'he began to study handwriting as a fine art. By practice he soon mastered it; and the texts of his painted books show a steady advance in skill of execution.' Mackail does not indicate whether Morris studied the art alone or with a practitioner. A page in the Bodleian Library on which Morris demonstrates a change in his handwriting after he 'went to Mr. Jones,' seems to apply to a script he used in his later calligraphy. In any event, Morris chose not to use any gothic lettering in the manuscripts of this period.

Illumination was a great delight to Morris the artist and lover of color. His daughter May remembered the brushes, the quill pens ('from a goose quill to a crow quill'), knives, rules, and compasses, as well as the colors and the gold leaf on his work table. She was especially impressed by the sure touch of the master craftsman.

> I have watched his firm broad hand as it covered a gold square half an inch in size with wee flowers formed of five pin-point dots of white laid with the extreme point of a full brush. The least wavering would have meant a jog or blot, but the blossoms grew with ease and surety.

When it came to dealing with gold, Morris tried various methods dating from the twelfth-century *De diversis artibus* of Theophilus Rugerus to contemporary practitioners. However, 'he chose his own method, and with much success,' May reports, thereby stating the essence of her father's practice in all fields, artistic, literary, typographic, or otherwise: study the masters, then do it your own way. May's insight into her father's attitude toward his work is worth repeating:

> It is a pretty thing to think of the illuminator's patient art blossoming under the hand of so impetuous and eager a man, showing a side of him not entirely recognized. My father was extraordinarily patient—with things if not with people always. There is the impatience of the trifler who hates his work, and that of the worker who hates trifling; it is needless to say which of them his was, and he had of course the balancing patience in dealing with things as means to an end—with lapses.

Though Morris greatly preferred to write and paint on vellum, frequently he had to settle for paper. The only vellum he could obtain without the white lead preparation which injured the colors and the gold was produced in Rome. A number of letters to his friend and one-time associate Charles Fairfax Murray in Italy show the problems involved in getting the amount, the size, and the quality he desired.

Between 1869 and 1875 Morris turned out an astonishing amount of calligraphic and illuminated work for a man deeply involved in running a business, designing patterns, writing poetry, and translating sagas. Alfred Fairbank has estimated that Morris produced 'more than 1500 pages of careful writing in several styles, and a great deal of ornament.' The manuscripts range from single sheets and trial pages to two that exceed two hundred pages. Only a few were entirely completed. In some he finished the text, but not the ornamentation, and in many neither was completed. He took his texts mostly from his translations of the Icelandic sagas; but three copies of Fitzgerald's first version of the *Rubaiyat* of Omar Khayyam and one book of his own verse were also among his major endeavors. His last two manuscripts in the Latin of Horace and Virgil, were the only ones in which he did not use the English language.

Though it is not dated, there is good reason to think that *The Story of the Dwellers at Eyr*, now at the Bodleian Library, is the first of his decorated manuscripts in this period. It is written in a hand that shows a relationship to the sans serif lettering seen on the margins of 'Bellerophon.' Morris shows here that he had not yet developed a feeling for the flat-nib pen. For most calligraphic effects he relied on curved ascenders in his lowercase letters and on adding curls, curves, and twists to his capitals as any enthusiastic amateur might have done. What his capitals lack in style they make up in verve, but the effect is not entirely happy when they are assembled into words. By the time he reached page 21, Morris seems to have felt a change was needed, and while he continued to use swash capitals in the text, he employed more conventional ones in the all-capital words that introduce each chapter. Pages 37–46 are missing, but on pages 47–50, as the manuscript ends, the thin diagonal strokes mentioned earlier, on h, k, m, and n, suddenly appear. They may also be seen in the small script he used for the chapter headings, although these were added later. Also added later were headlines beginning with page 21, in roman letters with serifs and tall ascenders.

On the first page, between the title and the text, Morris placed two rows of leafy sprays, and to introduce the story he painted a gold initial on a background of dark green reeds and red and white blossoms. From the corners of the square containing the initial, narrow feathery decorations grow out in several directions. Initials on colorful backgrounds with floral accompaniment introduce succeeding chapters.

Though he varied them greatly, the chief elements of Morris's manuscripts are shown here. From the first, his pages show his vitality, his mastery of color, his delight in the decorative qualities of natural growths, his openness to new ideas, his lack of imitation, and his pleasure as an artist in telling a tale with handsome letters and adorning it with living verdure.

In both its calligraphy and its ornament, *The Story of the Volsungs and Niblungs* follows the incomplete *Dwellers at Eyr*. While this work is also unfinished, Morris wrote about five-sixths of it before he stopped. Morris considered this saga the great epic of the people of northern Europe which could take a place beside the *Iliad*. In 1870 he published his translation of it, and he decorated the first page of a few of the large-paper copies with rows of pink and blue flowers separated by tufts of grass. The first page of the manuscript shows similar treatment with regularly placed flowers filling the margin around the opening lines of the saga. They were never colored, however, nor were seven of the eight medieval girls with musical instruments who stand among them. The girl in the upper right corner wears a long red robe and displays cymbals. Near her, in place of an initial, is a miniature by Fairfax Murray of Sigurd sitting on the lately slain dragon, Fafnir. Though Morris colored the initials for perhaps a dozen chapters, and drew rows of blossoms between the end of one chapter and the opening of the next, most of the decoration was never completed. The lettering at the beginning of the saga resembles the script in the latter pages of the *Dwellers at Eyr*. Most of the capital letters gradually lose curls and excess strokes to become conventionally roman in the course of time.

It is probable that Morris abandoned work on the *Volsung* manuscript when he decided to devote more time to writing and decorating a collection of twenty-five of his own poems for Georgiana Burne-Jones. At this period in his life, when his wife's affair with Rossetti was at its height, Morris found reassurance in his friendship with 'Georgie,' who had marital problems of her own, and for her he completed four illuminated manuscripts. The title of the first, *A Book of Verse*, takes on special meaning when one realizes that in the earliest version of Fitzgerald's translation of the *Rubaiyat* (which Morris wrote out later for Georgie), the familiar stanza reads:

>Here with a Loaf of Bread beneath the Bough,
>A Flask of Wine, a Book of Verse—and Thou
>>Beside me singing in the Wilderness—
>And Wilderness is Paradise enow.

Morris worked on *A Book of Verse* between February and August 1870, and brought into the project Burne-Jones who painted the first illustration, Fairfax Murray who painted the rest, and George Wardle, manager of Morris and Co., who drew the ornaments on the first ten pages and 'did all the colored letters both big and little,' according to Morris's colophon. Apart from the writing, Morris colored Wardle's drawings, did the ornament from page 11 to the end, and drew the minstrel figures on the title page, and the figures of Spring, Summer, and Autumn for a poem on the seasons to which Murray added the color. Despite the pensive melancholy that pervades Morris's verse, the foliage, fruit, flowers, and vines are full of the green vitality of spring. The entrance to this world is by way of a title page almost entirely covered by green leaves. Gold capitals set in the foliage give the requisite information; a miniature of Morris painted by Murray is placed in the middle of the page, and below this we see again medieval maidens with musical instruments—this time standing by a fence beneath some trees. The green growths pervade all the pages of the book.

Morris had not attempted any overall ornamentation before, but in *A Book of Verse* he had a smaller format and fewer pages to deal with than heretofore. Furthermore he restricted his ornament to the available space within a rectangle on each page, which it shared with the poetry and here and there a miniature. In his *Volsung* manuscript, whenever a song reduced the length of the lines, Morris filled the resulting space with sprays of leaves and blossoms. In *A Book of Verse*, however, the decoration tends to be of primary concern and it is not unusual for words at the end of the longer lines to burrow into it. This primacy of nature over letters occurs again in later manuscripts.

On page 11, when Morris took over the design of the ornament, a fundamental change becomes visible. The formal lines of sprays or small flowers disappear (except on pages 26 and 27) and are superseded by unified growths that rise from the base of the rectangle, filling the area to the right of the poetry, and reaching between the stanzas. From this time on, Morris decorated his pages, whether painted or printed, with growing, expanding foliage, flowers, and fruit rather than with formal rows of little plants or cut flowers and repetitive sprays. But though the ornament ceased to be static and became quite 'active,' it did not intrude into the margin, except on special occasions, in the manuscripts where it was confined with the text by rectangular rules.

Morris wrote *A Book of Verse* in the same hand used for *The Story of the Volsungs and Niblungs*, but to a few of the capitals he gave more than one form. Burne-Jones's miniature at the beginning of the first poem, showing a young couple divided by a stream, is somewhat heavy for the writing and the verdure below it. Fairfax Murray's illustrations vary in shape and size: five are roundels which lie like colorful islands in a sea of green foliage, and in three others he uses panels joined together to illustrate the adjacent poems.

Mackail, in commenting on the manuscript, said: 'Here there is a modernity which owes nothing to any tradition: and a freshness, a direct appeal to first principles and instincts which ... charms by its simplicity and fitness.' More recently Hermann Zapf has pointed out the strength and uniformity attained by Morris and his friends in their cooperative endeavor.

Morris's next gift to Georgie, however, was done entirely by himself except for the laying on of gold leaf in three places. It is *The Story of the Dwellers at Eyr* (Birmingham Museum & Art Gallery), complete this time with a Prologue, an Epilogue, and an index. It was finished in April 1871. It consists of 254 pages and is striking evidence of his enormous industry. Both the lettering and the decoration show that Morris borrowed for this edition over one hundred pages already written for the earlier *Eyr* manuscript. He rewrote about forty pages at the beginning and then completed the tale with a little over a hundred more. The chief difference between the scripts used in the two *Eyr* manuscripts is the addition of a serif that slants left from the top of the ascenders that curve to the right. The thin diagonal strokes are very much in evidence. As he rewrote the earlier pages, Morris revised the wording of his translation here and there.

Though Morris made no title page, he decorated the first page with a mass of stems, fronds, blossoms, fruit, and acorns, which fills the margins and is barely restrained from overwhelming the text by a frame of blue poles. As in other instances of his overall decoration, Morris uses two basic forms and shades of leaves which interact over the surface. The arrangement of the writing on this page is the first instance of his introducing the text with massed capital letters. They are not only to be seen on a number of later manuscripts, but are frequently used to introduce the poetry or prose printed at the Kelmscott Press. On the succeeding pages colored initials on colorful, often flowery, backgrounds precede each chapter, and they are accompanied by marginal decoration. On the pages taken from the earlier *Eyr* manuscript this decoration tends to stay close to the initial or to the side of the text, whereas the flowerings on the later pages take all the space they please. The short lines of the frequent songs are now placed in the middle of the page with vines and sprays on either side.

All in all, this manuscript shows how much invention, skill, and endurance went into its creation, and suggests the strength of the motivation that caused him to carry it through. Yet before he had finished the *Eyr* saga, he began work on still another gift for Georgie: the *Rubaiyat* of Omar Khayyam.

When Morris began to write out the *Rubaiyat* early in 1871, he had mastered a roman script distinct from that in which the foregoing manuscripts were written. One may see Morris changing from the latter toward the former in a pair of sagas, *The Story of Hen Thorir* and *The Story of the Banded Men* (Bodleian Library), for which he finished the text but not the ornament. They are representative of a group of manuscripts for which Morris used a two-column format. They are introduced with the title set in a block of capital letters at the head of the left-hand column, and except for the first pages, the decorations, which usually occur between chapters, stay within the bounds of their columns. The first ten pages of *Hen Thorir* are written in the script of the later *Eyr*. On page 11, however, the diagonal strokes disappear, the ascenders become straighter and the serif to the left of the vertical becomes shorter. Morris continued to employ this script through the opening pages of *The Banded Men*, but from there on he evidently made the manuscript a proving ground for variant forms of certain letters. He seems to have been more interested in variety than in consistency at this point, and it is possible that he worked on it at widely spaced intervals.

Allied to these manuscripts by their format and the use of roman script are *The Story of Kormak the Son of Ogmund* and trial pages for *The Story of the Ynglings*, 'Hafbur and Signy,' and *The Story of Frithiof the Bold*, all in the Pierpont Morgan Library. They show less steadiness in the formation of roman letters than he subsequently achieved. The use of tall ascenders, two or three times the x-height, may be a reminiscence of Arrighi's 'Lettera antiqua.' Morris completed the text of the *Kormak* saga and drew much decoration but did not color it. There is reason to think that he may have rewritten the first two pages. His control of the pen is better on them than on the immediately succeeding pages, and his treatment of the serifs at the top of the ascenders resembles the way he handles them in his later pages.

The only manuscript that shows the full effect Morris intended to give with a double-column format and its decoration is *The Story of Frithiof the Bold* (Doheny Library) for which he finished the writing but had only begun to color the decoration when he gave it to Fairfax Murray, who, with the help of Louise Powell, completed the ornamentation. Masses of green foliage with red and blue blossoms swirl between the chapters and around the songs. Murray carefully noted that Morris drew and colored the decoration for columns 3, 4, 5, 6, 8, and 11, plus some drawn headings to three others, and he adds, 'W. Morris set the pattern of the initials to the 2nd, 3rd and 4th chapters.' The script is similar to the roman used in the trial sheets.

Though Mackail states that Morris began work on the *Rubaiyat* before he finished the *Eyr* saga in 1871, it seems likely that he concentrated on it in 1872. This was the year that Georgiana Burne-Jones recalled that he worked on it, and in October of that year he gave it to her. He had borrowed for his purpose the copy given by Swinburne to Burne-Jones which contained Fitzgerald's first version. On the twenty-three vellum pages of this six-inch book Morris lavished some of his richest and most intense ornaments. On most of them foliage, flowers, fruit, and vines aggressively crowd about the small roman letters in the central rectangle. Though the format is that of *A Book of Verse*, the springlike tints of the earlier book have changed to the richer colors of summer with its recognizable growths. On five pages Morris shifts the emphasis to the margins with a great display of gold and green leafage, leaving the poetry unencumbered. An inner border restrains the twining vines much as it did on the first page of the second *Eyr* manuscript, and the presence amid the foliage of female musicians in long robes has by this time become familiar. At the head of each of these specially decorated pages a capital G quite clearly designates the recipient. The pages so treated are the first page; the two-page opening where the gilt words KUZA NAMA, between stanzas 58 and 59, are nearly obscured by a dense green thicket; and the last two pages of the poem which ends with two figures designed by Morris, surrounded by willow leaves, holding a scroll bearing the words TAMAM SHUD. At least one of these figures appeared first as an angel on the ceiling of Jesus College Chapel, Cambridge, and in 1890 as one of four figures in Morris's tapestry, 'The Orchard.' Burne-Jones assisted Morris in designing the thirty-one full and half-length figures in the margins, and all were painted by Murray. The gilt flowers on the red morocco binding were stamped with tools designed by Morris; the effect anticipates that of Cobden-Sanderson's bindings in the 1880s.

May Morris called this *Rubaiyat* 'a jewel-like book.' Mackail felt that its 'treatment of the fruit and flower work is an admirable adaption of an almost Pre-Raphaelite naturalism to the methods and limits of ornamental design.' Vallance agreed that the naturalness is at the same time perfectly decorative, and asserted that it carries on the art 'to a stage of evolution of style in advance of anything it has ever attained before.'

A second *Rubaiyat*, made for Edward Burne-Jones, was worked on in 1872 and 1873, according to Mackail and May Morris, which seems more likely than Graily Hewitt's date of 1871. Though planned along the lines of its predecessor, the foliage tends to be darker and less in quantity, and the blossoms and fruit more pallid. No pages shine with gold leaves in the margins. The writing is similar but the ascenders are taller. Though more restrained in style, this *Rubaiyat* is unique in having six miniatures by Burne-Jones. One is on the title page above the title set in green foliage. The rest are placed at the heads of their respective pages, and resemble in size, colors, and characteristics the picture he contributed to *A Book of Verse*. Most of them depict affectionate couples, and their connection with the verses is not always apparent. For some reason, in this manuscript Morris spelled the author's name with one Y rather than with the customary two.

In January and February 1873 Morris was translating *The Heimskringla*, 'The Stories of the Kings of Norway Called the Round World,' by the thirteenth-century

Icelandic historian Snorri Sturluson, which he published some twenty years later in *The Saga Library*. Snorri's Preface and the first section, *The Story of the Ynglings*, were written out calligraphically on 109 pages of vellum. Morris begins with a script that has some similarity to that of *Hen Thorir*, but soon it becomes more regular as a variation of his high-ascender roman hand. The serifs on the ascenders slope across the top of the vertical stroke, the lowercase a and e appear frequently in an uncial form, the t has an awkward flick added above the crossbar, and exaggerated serifs end the horizontal strokes of the capitals E, F, and T. This is one of the scripts Morris tried in *The Banded Men*. Though the writing here may be less attractive than in some other manuscripts, the decoration, so far as it was done, is as colorful and inventive as ever. The Preface is introduced by dark letters on a band of willows, and by a white initial on a background of green leaves and white florets. As before, Morris ornamented the chapter openings with such initials and with flowers and foliage in the margins. Here, however, the familiar young women in robes, some of whom hold the expected musical instruments, may be seen again. The few of these figures that were actually colored are a decided addition to the marginal ornament. Drawn in sepia by Philip Webb on page 4, two lone rabbits await the greenery that never came. On page 15, Fairfax Murray painted a miniature of the head of Odin whose exploits are told in the early chapters of the *Ynglings*. In several places Morris painted spiraling acanthus on a background of the customary slender green leaves. They are better developed than those he had painted earlier, but they were still sharing space with the forms they would eventually supplant.

The Story of Halfdan the Black and *The Story of Harald Fairhair* follow *The Story of the Ynglings* in *The Heimskringla*. In his calligraphic manuscripts of these two texts, Morris continued to use the same script, but in the *Harald* story he introduced some variations in the capital letters. One, in which the capital V gains a sinuous left stroke, is directly related to the V of *Three Icelandic Sagas* (see below). That these sections were to have been decorated is shown by the blanks left at the beginning of each chapter. The title, the heading for chapter 1, the initial, and first word of *Halfdan* were lettered in brown ink, as was done with a space filler of stylized fronds reminiscent of *Hen Thorir* and *The Banded Men*. These elements would doubtless have been colored had the original intentions been carried out. Morris's rendition of these early portions of Snorri's history has come to rest in three places: Preface and *Ynglings* at Kelmscott Manor, *Halfdan* and the first third of *Harald* at the Pierpont Morgan Library, and the balance of *Harald* at the Bodleian Library.

About 1873 Morris experimented with a sweeping slanted hand with long stiff ascenders and descenders. Two examples are illustrated by May Morris in the *Collected Works* of her father: *Lancelot du Lac* in volume IX, page xxi, and *Journal of Travel in Iceland* (which Morris finished fair copying in July 1873) in volume VIII, page 98. For *The Story of the Men of Weapon Firth* (Emery Walker Collection) Morris used a similar slanted script in its eighteen pages. Here the writing is more condensed and somewhat more italic than in the examples above. Swash capitals give Morris a chance at sweeping strokes of the pen. Some capitals, such as C, E, and L that dip below the letter line, closely resemble forms used by Arrighi and Tagliente. The S is kin to the form in *Three Icelandic Sagas*.

Two manuscripts which also have capitals with sweeping strokes and italic-oriented lowercase letters with long ascenders and descenders are *The Tale of Haldor*, of which three pages exist, and *The Story of Egil Son of Scaldgrim*, with 157 pages, now at Kelmscott Manor. On these pages, however, the axis shifts from the diagonal to the vertical even though long diagonal serifs surmount the ascenders. Since both ascenders and descenders intermingle and nearly touch the lines above and below them, the page appears to be woven. Here again the influence of the Italian writing masters is evident. At least nine capital letters display characteristics to be seen in their books.

On page 125 of *Egil* the writing becomes distinctly heavier and rougher. The nib of the pen is broader and the distribution of ink is often uneven. The lowercase g is particularly crude, while the curves of capital B and R are made up of thick and thin strokes that substitute angles for curves. Morris improved his pen by page 144, and in the final pages one can see developing the hand he used in *Three Icelandic Sagas*.

Except for the first page of *Egil*, the foliage at the chapter heads is colored pen-work. The initial is not enclosed in a separate square, and the whole decorative effect is much lighter than heretofore. In the lettering of chapter titles, beginning with Chapter VII, one sees the appearance of flagella extending from various ascenders, particularly from f, which may entangle adjacent letters or tie themselves in knots. Decoration on the first page was not completed but it shows a capital T involved in white vines which spread out into the margin and are accompanied by many little discs with pen-drawn rays around them—of which more hereafter.

In several ways the *Egil* manuscript is linked by its characteristics to certain other manuscripts of the 1873–74 period. On 26 March 1874 Morris wrote to Murray that he had 'taken rather to the Italian work of about 1450 for a type—this kind of thing don't you know.' To show what he meant, Morris sketched a capital A behind which were four coiling vines with a toothed flower growing from one of them. Strikingly like this sketch is the initial with which he introduced a third *Rubaiyat* (Bodleian Library). The text was written in two varieties of small roman script, and very little ornamentation was done beyond the first page. There, however, the large golden A in a red and gold square, entangled by curling vines with their blossoms, dominates the page—all the more so because the massed capitals of the first line and a half of the poem stand to the right of it, while mid-fifteenth-century vines sprouting toothed flowers and early fifteenth-century 'sunbursts'—the gold discs with rays—pervade the left-hand margin. In this elaborate ornamentation the initial A is less distinct than the white and active vines that surround and partly cover it.

Morris's use of a heavy script in *Egil* may have been a deliberate attempt to try the kind of writing taught by a Mr. Jones. In the Bodleian Library, as we have seen, there is preserved a single sheet of paper, undated, on which Morris wrote in straggling longhand: 'This is my hand-writing before I went to Mr. Jones and this afterwards.' Below, he wrote, 'what do you think of this change of hand-writing' in heavy, slanted but better-formed letters. Both sentences were hastily written. In the latter, ascenders curve to the right, the lowercase g is particularly rough, and the f has a descender that curves to the left. Though it is cursive writing, it is related both to the calligraphy we have seen in *Egil* and also to that used in a manuscript of nine pages in which Morris recounted two stories and part of a third one relating to Jaffier, at one time a favorite of Haroon al Rasheed. The text comes from *Tales, Anecdotes and Letters, translated from the Arabic and Persian*, by Jonathan Scott, published in 1800. In this dark italic hand the ascenders curve to the right, the descender of the f curves to the left and the w resembles the form presumably learned from Mr. Jones. Even the g, much smoother here, may be recognized. Among the capitals which exhibit sixteenth-century Italian characteristics, one meets the N with the curved descender. If it were not for the watermark date of 1872 and other indications of its chronological position, one might assume from these characteristics that it was contemporary with the earliest group of manuscripts, but written with a different pen.

Though the text ends abruptly in mid-sentence, decoration was begun and some of it was painted. The first word, in red roman capitals, follows a red initial W behind which white vines have been drawn. The marginal decoration is made up of abstract pen-work in the medieval tradition rather than by recognizable natural growths.

Red pen strokes run vertically from the upper margin most of the way down the left side of the text, and horizontally from the top of the W to the left side of the sheet, forming a kind of red cross. The initial T on the final page is surrounded by uncolored acanthus leaves. It might almost be an initial designed for the Kelmscott Press. Its lack of a square to define a background relates it to Morris's designs for initials in the later Kelmscott volumes.

A very interesting page, now owned by Mr. Norman Strouse, brings together the scripts which Morris was trying out at this time. One is very much like the slanted hands mentioned above which culminate in *The Men of Weapon Firth*. Another set of lines resembles the script in which he wrote the *Aeneid*. The third is the heavy italic, just described, at a point somewhere between the 'Jones' note and the *Haroon* hand. He had not mastered it, however. In one place he refers to 'the thick pen if I could get it to go,' which indicates both his instrument and his struggle with it. Equally as revealing in another way is his attempt to reproduce with a broad-nib pen the elegant title page (without its elaborate initial) of Arrighi's writing book of 1523. The contrast is understandably great, but it is of interest to see how he attempts to imitate every feature of the capital letters. On the reverse side of the sheet is a trial page for the *Aeneid* which points to a date in the latter part of 1874, though some of the hands shown here may have been used in 1873, and there is no assurance that he wrote on both sides of the sheet at the same time.

On 26 March 1874 Morris informed Murray: 'I wrote a book (on paper confound it) of about 250 pp. translations of unpublished Icelandic stories with pretty letters to each chapter, which look well on the whole. I finished this early in February.' The book is *Three Icelandic Sagas* (Fitzwilliam Library) comprising *The Story of Hen Thorir*, *The Story of the Banded-Men*, and *The Story of Haward the Halt*, on 244 pages. It was his fourth gift of an illuminated manuscript to Georgie Burne-Jones, and her initials appear amid sprays in the lower margin of the first page. Considering the length of the work, it is likely that most of the writing was done in 1873 and the ornaments mainly in 1874. The script is an elaborate dark italic with serifed ascenders more moderate in height than in *Egil*, but with a number of the unusual capitals seen there (e.g., B, R, and V, described above). Some capitals derive from Lombardic examples, others he may have invented. He saw no need for consistency, and as with the Kelmscott Press initials, one finds several versions of the same letter. His wish to be elaborate in this manuscript caused him to use many hairline flourishes, notably a whip on the lowercase f which became more complicated as he went on. The headlines on each page, written in brown ink, have taller ascenders, larger capitals, and more flourishes than the text does. The chapter titles, written with a narrower pen, contain even more tangles of hair lines. Morris's most unique letters, however, appear to have no immediate ancestry, though some of them hint at uncial, Lombardic, or other forebears. They occur in the first word of each chapter following the initial, and at the beginning of lines of poetry. They are constructed with a certain regular irregularity that causes them to look as though they had been folded out of paper or fabric. This is the only manuscript in which they appear. Indeed, *Three Icelandic Sagas* is the only one of any length in which the calligraphy is quite so consciously ornate.

As in his other longer works, the decoration tends to cluster about the chapter openings. Each of the three sagas begins with a large roman initial tangled in vines on a colored background. Though Morris had previously allowed vines and foliage to decorate the margin, here the background accompanies the growths to form an irregular pattern to the left of the initial. Though the initial area thus assumes a free-form appearance, at the same time the growing elements are more contained than they were without the colored ground. In the first two sagas the vines embracing the initials at the beginning of the chapters are white or light in color; thereafter darker, richer colors prevail as in the *Odes* of Horace. The third saga allows these vines with the smaller initials to venture into the margin, whereas in the first two, except for Chapter 1, they stay within their squares like the wood-engraved initials of the Kelmscott Press.

Though Morris wrote his next manuscript, the *Odes* of Horace, in a condensed italic basically similar to the script he used in *Three Icelandic Sagas*, it was a more straightforward letter, and roman initials took the place of the variety seen there. The combination of Latin text, italic lowercase, and roman capitals suggests the Renaissance of Aldus rather than the classic Rome of the author. Using a crow quill pen, Morris completed writing the four books in 183 pages that were six and a half inches high. He did not, however, finish the decoration, which is not surprising since he intended to decorate the beginning of every ode. As he wrote to Murray in March 1874, 'The odes are short so there is nearly an ornamental letter to every page, which makes it a heavyish piece of work.' Even so, he did ninety-seven initials, but toward the end he was sketching the letters and ornaments in brown ink.

In this manuscript, as with the first *Rubaiyat*, Morris emphasized richness in a small format. His pages are interspersed with gold, silver, red, and blue capital letters in the text. Each *carmen* begins with an initial on richly colored backgrounds amid white and pink vines that flower in the margin. The left-hand margin, and sometimes the upper and lower margins as well, is alive with sunbursts lightly connected with pen-work, as was done for the third *Rubaiyat*. The effect of the fully decorated pages deserves May Morris's remark that they are '. . . instinct with joy, vivid and jewel like.' Before he ceased to color the decoration he had drawn, Morris initiated the use of pen-work that adheres to the ruled lines, without adding flowers and discs, in the way he ornamented *Haroon*. In both kinds of ornament, Morris was drawing more clearly on specific medieval decorative elements than he had done hitherto.

Morris intended to give especially colorful treatment to the opening page of each of the four books that make up the *Odes* of Horace. He completed only the first page of Liber II which clearly shows what he had in mind. A large silver initial among swirling golden vines introduces seven lines of massed capital letters (gold and silver) which overwhelm the small script below. In the margins billowing blue and green acanthus leaves coil about golden staves, and faces peer out of gold frames in each corner. On the first page of Liber III the faces are painted but the acanthus, though present, is not. Faces were painted for Liber I and Liber IV also, but no other decoration was done. A long time afterwards Morris wrote of these pages: 'There are some heads in the ornament designed by Burne-Jones and some of these are painted by C. F. Murray. I did those in the ornament of book 2.'

Though Morris may have felt that some of his work of decoration was heavy, he also found joy in his work. 'I am very keen on the thing just now,' he wrote in March 1874, 'and really enjoy my Sundays very much at it. . . . I was 40 years old last Monday . . . staying home and working hard at illuminating.'

About this time Morris thought of doing a manuscript of 'The Story of Cupid and Psyche' with illustrations based on the designs of Burne-Jones which were cut for the projected special edition of *The Earthly Paradise*. He resolved instead to create a masterpiece by writing out and illuminating all twelve books of the *Aeneid*. Late in 1874 Burne-Jones wrote: 'Every Sunday morning you may think of Morris and me together—he reads a book to me and I make drawings for a big Virgil he is writing—it is to be wonderful and put an end to printing.' Though he wrote out 177 pages, about half of the total, and did not get very far with the decoration, the *Aeneid* is a monumental work. May Morris thought it 'among the notable manuscripts of the world,' and Mackail felt it was by far Morris's best. The letters appropriately are roman, but differ greatly from his earlier use of the script in that they are wider, almost square in many instances, with the upper curves flattened on a, m, n, etc. Low ascenders no longer assert a visual dominance over the other letters. A number of such features, taken together, convey the impression of well-constructed masonry. Indeed, his banishment of decorated initials to the margins shows how important he felt this effect to be. His writing is uncrowded and regular. It has a classic breadth and serenity as it stands at the culmination of his practice of an art that began with the uncertain gothicism of *Paracelsus* and *Guendolen*.

Though dignified roman capitals stand at the head of his pages—blue capitals on the verso, gold capitals on the recto—and the same script is used for colored capitals in the text, he could not maintain his classic pose when it came to the large initials in the margins. They differ from any of Morris's previous designs for initials. His daughter reports that he liked to begin his decoration 'well on in the book, for he wanted the first pages to be as good as the rest, to be done when he had settled into the work and was at his best.' Though this had not always been his practice, in the *Aeneid* his decorative initials are found in the margins from page 43 to page 72, most of them in color. These initials are richly ornamented and not at all classical. Some have formal backgrounds, some have none, and some that have them wander out of them. No white vines have survived, and slender leaves are rare, but acanthus is ubiquitous whether in a background, as an entangling element, as an outgrowth of a letter, or as the substance of the letter itself. In this new form Morris was as inventive as ever.

A number of very large initials had been designed. One of them, a Q made of leaves, surrounds a figure painted by Murray. Burne-Jones designed a number of scenes to be painted in the margins, and half-page illustrations to come at the beginning of each of the twelve books. From his friend's design Morris painted the picture of Venus and Aeneas for the first book but he invited Murray to go over it as he distrusted his own ability to do this kind of work. As it stands, only the head of Aeneas is the untouched work of Morris. Below the picture, the opening lines of the epic in capital letters are inscribed in a rectangle that recalls an ancient Roman tablet. For these letters he laid the 'red bole ground ready for gilding,' according to his daughter, but he never got around to applying the gold. Eventually Morris sold the unfinished manuscript to Fairfax Murray who commissioned Graily Hewitt to write the missing six books. Murray painted the designs by Burne-Jones for the remaining illustrations and miniatures, while Louise Powell added borders to the opening pages of the books. Hewitt gilded the grape border she designed for the first page. In passing, one may notice that Aeneas has the familiar elongated Burne-Jones figure, so recognizable in his later painting, whereas the persons in the *Rubaiyat* miniatures appear much more normal in their proportions.

In 1875 Morris gave up calligraphy and manuscript illumination for lack of time: he had undertaken to translate the *Aeneid* into his own kind of verse, and he had added to his labors for Morris and Co. the study of dyeing to improve the color of the firm's woven goods. One may say that the vigorous acanthus now triumphed over the less assertive willow: Morris removed Rossetti from his house, reorganized his firm, and went on to take a prominent place among the public men of the day in art and socialism.

In 1888 Morris showed several of his manuscripts at the Arts and Crafts Exhibition Society's first exhibition. After he had seen the first page of the second *Dwellers at Eyr*, Bernard Shaw, a good friend of Morris, wrote in *The World*: 'The decorative diagonal strokes suggest the conventional representation of a shower, as affected by artists who design advertisements for macintoshes.' Only once did Morris return to calligraphy. About 1890 he began a catalogue of his library with sixteen pages of entries and comments written in a somewhat uneven italic. He planned to add decorated initials to the entries for his incunabula, and four of them were done with sprays behind the letters. Morris was designing his Golden type at the same time, was having his translation of the *Gunnlaug Saga* printed at the Chiswick Press in imitation Caxton type, and was writing *News from Nowhere* in which Dick says, in relation to 'handsome writing,' that in the twenty-first century 'many people will write their books out.'

In the following year when the Kelmscott Press was in operation, Morris wrote with a touch of nostalgia:

> Pleased as I am with my printing, when I saw my two men at work on the press yesterday with their sticky printers' ink, I couldn't help lamenting the simplicity of the scribe and his desk, and his black ink and blue and red ink, and I almost felt ashamed of my press after all.

In November 1894, when his printing was in full swing, he could take time to advise a correspondent on his calligraphy by drawing a diagram to illustrate how to cut a pen nib and suggesting that he seek models in 'some fine book written about 1120 or even earlier.'

Though Morris did not live to see the revival of calligraphy under Edward Johnston, he influenced the younger man through Sydney Cockerell who gave him advice based on close association with Morris, his library, and the medieval manuscripts he was collecting. Morris's own calligraphy does not attain the high quality of the art reached before and after his time, but by a study of it one sees Morris at work and feels his urgency to tell the story under his hand with special lettering and in a special style. One can watch his pen try various forms of letters as new ideas for different effects strike him. It is tantalizing—and fruitless—to speculate where his maturing craftsmanship might have led him. His later manuscripts show his ability to contrive fresh ways of handling familiar elements and his ability to go successfully in new directions. His experience in the art of combining all elements to present an aesthetically pleasing appearance on the page stood him in good stead when he was faced with the typography of the books of the Kelmscott Press.

Though his calligraphic impetus ceased, his decorative talent, which had made itself known before it was applied to manuscripts, continued for many years afterwards. One can recognize aspects of the works we have surveyed in some of his wallpapers, his fabrics, and his tapestries. The roof and roof supports of the library in the Oxford Union Society building (redecorated in 1875), the borders cut on wood in 1872 for a fine edition of *Love Is Enough* that was never published, the decoration on the binding of the published *Love Is Enough*, and the pages of the books printed at the Kelmscott Press all show a distinct relationship to the abundant decoration that Morris lavished on his manuscripts that very few people would ever see. To complain, as some people do, of the decoration given by Morris to his painted and printed books is to complain of a basic quality in Morris himself for whom decorative design was fully as natural as breathing and as essential to life.

The sources of Morris's manuscripts lie in the Middle Ages, the Renaissance, and the fruits of the earth. Whatever he took for his pages—script, decorative motifs, foliage, flowers—he made his own with overflowing originality. He used no gothic lettering in this period, but chose to work in the tradition of the humanist manuscript which took its capital letters from Rome and its lowercase letters from the days of Charlemagne. For half a dozen years he spent long hours of intense artistic creativity with pen and brush which demanded exceptional concentration, clearness of eye, and steadiness of hand. His hope to achieve a masterpiece had to be deferred until the completion of the Kelmscott Chaucer, but what he did accomplish is nonetheless remarkable. Few people have seen these manuscripts, fewer have written about them, and fewer still in the twentieth century have been able to comment more than superficially upon them. But two outstanding English calligraphers, Graily Hewitt and Alfred Fairbank, have discussed Morris's manuscript work with special insight, based on their own mastery of this ancient art. In his talk to the Double Crown Club in 1934, Hewitt summed up Morris's calligraphic achievement as follows:

> With the manuscripts one is at once aware, despite certain roughnesses, . . . of a grasp, a largeness, a lavishness, and delight in them, as in every enterprise he undertook, for admiration of which there seems no words. Everything is on the grand scale, regardless of time, trouble, or money to be spent on them. . . . These manuscripts show little of the Gothic tendencies of later book work; but the Renaissance methods are fresh as all his work is fresh, with an originality which probed to the essential bottom of the business and then burst upwards again with irrepressible vitality.

William Morris: Typographer

BY A TYPOGRAPHER,' WROTE JOSEPH MOXON IN 1683, 'I mean such a one who by his own Judgement, from solid reasoning within himself, can either perform, or direct others to perform, all the Handy-works and Physical Operations relating to Typographie.'

William Morris reached his mid-fifties before he qualified to meet Moxon's definition; but while still in his twenties he had developed a remarkably fine judgment from solid reasoning within himself that enabled him to perform, and also to direct others in his employment to perform, a wide variety of crafts and physical operations. Quite early in his life he had recourse to professional printers. At the age of twenty-two he edited the first number of the *Oxford and Cambridge Magazine*, published in 1856. He was fortunate to be served by the Chiswick Press, where a lively interest in typographical design had been fostered by the Charles Whittinghams, uncle and nephew, of whom more will be said shortly. Between 1856 and 1890 Morris was involved on at least eight occasions with the Chiswick Press over the printing of books, and also of trial pages for various projects of his that were never completed. But until he reached his mid-fifties, he appears to have lacked the time and inclination, and also the technical expertise, that he needed to practice typography with the same mastery that he displayed in so many other crafts. Yet it was precisely that period of creative activity in other fields that equipped him so well to become a typographer. By the time he finally embarked upon type design, he was able to draw upon his practical experience of calligraphy, and also of designing and printing wallpapers and furnishing materials. He had also learnt how to obtain the right ingredients and processes to make dyes which yielded the superb effects demanded by his own unbending standards. Consequently he had a very sound basis on which to decide matters to do with the practice of typography.

Before he decided to design his own types, Morris had become involved in the typographical arrangement of books made from his own texts, even though they were at that time composed in types designed by earlier craftsmen. The resources of the Chiswick Press included founts of Caslon, the pioneer eighteenth-century English trade typefounder, whose types were quite acceptable to Morris. In addition there was available at the Chiswick Press a roman type named Basel, recut in the style of type design that was often found in early sixteenth-century books printed at Basel. Morris had a predilection for this type: he used it c. 1867 to set specimen pages for *Cupid and Psyche*, for which alternative settings were made in Caslon type. The Basel type was used again later for his prose romance *A Tale of the House of Wolfings* (1889) and for another of his romances *The Roots of the Mountains* (1890). Charles Whittingham had ordered the type from a small typefoundry in London run by William Howard from 1838 to 1859. From the same supplier Whittingham obtained a replica of one of Caxton's *bastarda* types. This too was employed by Morris for his translation from the Icelandic of *The Story of Gunnlaug*; but the setting was never used for a published edition, and only a few copies were circulated in 1890. An earlier abortive typographical project is known from two trial leaves produced c. 1872 for *Love Is Enough*. They are of interest mainly because of the initials and marginal decorations, one from a design by Burne-Jones, which Morris engraved on wood. Their effect was spoilt by combining them ineptly with a feeble 'modern' typeface, and it is doubtful whether these trials were printed at the Chiswick Press.

In addition to its interesting and unusual range of text types, the Chiswick Press offered a rich choice of decorative initials, many of them recut after sixteenth-century models by Mary Byfield (1795–1871), a talented member of a family of engravers, who devoted her life to the support and education of her many nephews and nieces. As Morris used some of her initials, he probably recognized Mary Byfield's unusual gifts for rendering both letters and pictorial illustration upon wood; but he was certainly aware that many other skilful trade wood-engravers were to be found at that time in England. What he did not know was how to design types, nor where to have them cut and cast.

Five years before Morris finally decided to make his own types, and to set up his own press, Kegan Paul had published in the *Fortnightly Review* an article on 'The Production and Life of Books' in which he wrote: 'there could scarcely be a better thing for the artistic future of books than that which might be done by some master of decorative art, like Mr. William Morris, and some great firm of typefounders in conjunction, would they design and produce some new type faces for our choicer printed books.' At the time those words were published in 1883, Morris was more interested in the use of printing to further the cause of socialism than in the creation of 'choicer printed books.' He had joined the Democratic Federation in January 1883, and thereupon 'speedily showed that he meant to work in grim earnest on the same level as the rank and file of our party,' as H. M. Hyndman remarked. Though Morris resigned in 1885 from what had by then been renamed the Social Democratic Federation, he at once took a leading part in the newly formed Socialist League, and became editor of its official organ, *The Commonweal*. Its foreman printer was Thomas Binning, later to become father of the chapel at the Kelmscott Press. In a letter to Andreas Scheu on 16 July 1885, Morris wrote of Binning: 'He is to be my man you understand, and I am to be the capitalist printer.' So strictly did he concentrate upon his dual role as editor and as capitalist printer that he even refrained from raising any objection when Binning chose to compose *The Commonweal* in a so-called modern style of type. In fact Morris disliked such forms, especially the 'sweltering hideousness of the Bodoni letter, the most illegible type that was ever cut, with its preposterous thicks and thins.' Meek acceptance of a modern-style typeface for composing *The Commonweal* simply reflected his preponderant interest at the time with the form of socialist doctrine, to the virtual exclusion of any concern with its typographical arrangement, which he left to Walter Crane and George Campfield for decoration and supervision.

During these years of vigorous political activity (and he was as active on the platform and street-corner as he was in print), he became closely involved with several friends and acquaintances who shared his concern for the arts and crafts as intensely as his socialist ardour. Three such friends were founder-members with Morris of the Arts and Crafts Exhibition Society in 1887. They were Emery Walker (1851–1933), T. J. Cobden-Sanderson (1845–1922), and Walter Crane (1845–1915). Each was later to contribute significantly to the achievements of the Kelmscott Press, but by far the most important role was to be played by Walker.

The idea of forming a new society to interest the public in the movement for raising standards in the useful arts was first mooted in 1886 by the architect W. A. S. Benson. He knew Morris well and worked for his firm. The scheme was one which Ruskin had commended to Morris on 3 December 1878 when he wrote to him, 'How much good might be done by the establishment of an exhibition, anywhere, in which the Right doing, instead of the Clever doing, of all that men know how to do, should be the test of acceptance.' In 1887 a committee was set up, but its first unattractive title 'Combined Arts' was changed to 'The Arts and Crafts Exhibition Society' at the suggestion of Cobden-Sanderson, who also argued that its first exhibition in 1888 would have to be the occasion for arranging a series of lectures on various crafts. He himself undertook to speak on bookbinding, and he succeeded in persuading his friend Walker, much against his will, to lecture on printing. The opening lecture was given by Morris on tapestry and carpet weaving; and the final lecture on the general topic of design was given by Crane.

From these brief details of the newly formed Society and its first exhibition, it will not be apparent why it exerted such a strong influence upon Morris's concern for the art of typography. The reason is twofold: first, because the mounting of the exhibition led him to realize that he had not yet produced a book in a typographical form of sufficient distinction to warrant its submission for exhibition; and next, because Walker's lecture on printing opened his eyes to the method by which he could embark on the task of creating his own types, as a first step in the establishment of his own private press.

The two earlier essays in this catalogue have made it clear that by 1888 he had already formed an exceptionally discriminating taste for books, notably by his intimate knowledge of the manuscripts and early printed books in his own library, and also by his skilful production of magnificent formal manuscripts and calligraphical trials. In this way he had developed a fine appreciation of letter forms, and of the right disposition of space to set them off to best advantage. At the same time he acquired a keen eye for quality in paper, ink and presswork.

His appreciation of the finer points of typography was considerably extended by frequent meetings with Emery Walker from 1884. Both men had houses in Hammersmith overlooking the Thames. Walker lived little more than a quarter of a mile upstream from Kelmscott House, where he often called on his way to or from his printing works. His works were installed in a fine eighteenth-century building named Sussex House which lay only a few doors downstream from Kelmscott House. His visits became so frequent and so welcome that Morris later declared that a day in Hammersmith was not complete without a visit from Walker.

A shared interest in socialism had brought the two men together. In 1885 Morris and Walker established the Hammersmith Branch of the Socialist League, for which Walker acted as secretary and Morris as treasurer (their branch membership card was decorated with a design by Walter Crane). Though the cause of socialism remained one of their many shared interests, it was Walker's practical and historical knowledge of printing, together with his gentle and obliging nature, that made him of such vital importance to the typographical activities of William Morris.

Walker was seventeen years younger than Morris and came from a very different background. Born in London in 1851, he was the eldest of five children of a coach-builder from Norfolk. Because his father went blind when young Emery was only thirteen, he had to leave school in order to earn money for his family's support. A more fortunate turn of events soon brought him under the influence of Henry Dawson, a kindly landscape painter who devoted much of his time to the Congregational Church in Hammersmith where Walker attended Sunday school. Dawson's elder son founded The Typographic Etching Company in 1872. A year later Walker joined the firm. Thus at the age of twenty-one he began to acquire that extensive practical knowledge of printing and of reproductive processes from which so many of his contemporaries were later to benefit.

The Typographic Etching Company was conducted by men who were artists at heart, and they claimed to be the only firm in London which could offer three different processes of reproduction—for making line- and later half-tone engravings, as well as two methods of photogravure reproduction. After two years with this company, Walker married, and at Christmas 1879 moved with his wife and daughter into a pleasant house on Hammersmith Terrace. In 1883, a year before he met Morris, Walker left The Typographic Etching Company to join his brother-in-law in a printselling business; but after three years he returned to the printing trade. He then set up in 1886 with a partner as Messrs. Walker & Boutall, Automatic and Photographic Engravers, with offices at 16 Cliffords Inn, off Fleet Street, and with his works in Hammersmith at Sussex House.

Despite his extensive experience of printing, Walker was most reluctant to lecture on the subject to the Arts and Crafts Exhibition Society in 1888 because he had no experience of public speaking, and suffered acutely from shyness and nervousness. He therefore wisely decided not to rely solely upon his uncertain gifts for verbal exposition, but instead to make use of his own company's facilities to produce lantern slides with which to illustrate his lecture.

In the late 1880s, photographic lantern slides were still a novelty; they certainly were of crucial value for Walker's lecture on printing at the New Gallery, where the Arts and Crafts Exhibition Society held its first show. Reporting on the lecture in the

Pall Mall Gazette, Oscar Wilde declared that nothing could have been better than Walker's lecture: 'a series of most interesting specimens of old books and manuscripts were displayed on the screen by means of the magic-lantern, and Mr. Walker's explanations were as clear and simple as his suggestions were admirable.'

Most of the old books and manuscripts projected onto the large sheet hung in the New Gallery would have been thoroughly familiar to Morris, for he had lent many of the originals from which Walker's slides were photographed. But what came as a dazzling surprise to Morris was to observe the qualities of such large photographic enlargements of fine fifteenth-century typefaces when they were projected onto the makeshift screen. After the lecture he became greatly excited for reasons which were recorded by his daughter in a description of his behavior at the time. He was struck by the appearance of the finely proportioned letters when enlarged to such an enormous size in Walker's slides, and also by the fact that they had gained rather than lost in the process.

The enlargements had the effect of emphasizing to Morris the qualities of the types. Characteristically he felt that if such results had been obtained in the past, they could be obtained again. He felt an overwhelming desire to try his hand at type design, and on the way home to Hammersmith from the lecture said to Walker, 'Let's make a new fount of type.' And that, according to May Morris, was how the Kelmscott Press came into being. It was the outcome of many talks between the two men, but it had needed this impetus, the spur of excitement, to turn the desirable thing into the thing to be done.

Those evanescent images upon the screen could of course be repeated in a somewhat smaller scale for leisurely study by Morris. To this end Walker supplied his friend with enlarged photographic prints made from the same fifteenth-century typefaces that had been projected as slides. Several of these prints were closely scrutinized, carefully traced, and then patiently redrawn by Morris—but not with any intention of producing exact copies to enable the original typefaces to be recut. He used the photographic enlargements as a means to unravel the secrets of how fifteenth-century craftsmen had applied their skills to the problems of type design.

As experienced designer, Morris was perfectly capable of using fine tools for engraving, but his impatient temperament and stocky physique led him to prefer working on a larger scale to obtain quicker results. A sketch by Burne-Jones of Morris making a wood-block for *The Earthly Paradise* shows him busy with his set of gravers:

half of those not actually in use are strewn around him on the floor. So Walker's enlargements provided him with a merciful release from working in a tiresomely small scale. Furthermore Walker was able to find him an exceptionally skilled craftsman who was able to cut punches in faithful conformity to Morris's original drawings. His name was Edward Prince (1846–1923) and he was later to cut punches for many other English and continental private presses; he had been apprenticed to another well-known punchcutter named Frederick Tarrant, who had worked for Walker while he was with The Typographic Etching Company.

Walker's trade connections proved to be as valuable to Morris as his technical knowledge and advice. It would nevertheless be wrong to consider Walker as just a friendly and useful technical expert. He had developed a deep love of books and a store of bibliographical knowledge that was remarkable in a man of his simple background. As a boy he read most of the books at home and then began to rummage through the penny and twopenny boxes outside booksellers' shops. He made a habit of buying anything in them which seized his attention and for which he had enough pennies. When he was about thirteen he came across an elementary book on bibliography, from which he first learned something about the invention of printing and the names of eminent printers. Armed with this knowledge he bought as many examples of their work as he could afford. He also made extensive use of the British Museum, where he studied both printed books and manuscripts. In addition, by learning to draw and by mastering the techniques by which illustrations could be reproduced, he acquired a fine taste for book illustration. He was therefore remarkably well qualified to assist Morris at the Kelmscott Press, despite the inequalities of their ages and their creative talents.

Walker also had the necessary commercial experience to be able to set Morris's mind at rest at the outset of his typographical adventure on the matter of printing costs. Though Morris was in comfortable circumstances, from the time of the New Gallery lecture until November 1890 he was paying subsidies for the production of *The Commonweal* to the tune of several hundred pounds a year. Walker was able to produce detailed estimates to show that Morris could hope to produce and enjoy a 'decent-seeming' book in a sufficient quantity to distribute copies among a few chosen friends for about the same amount of money that it would then have cost him to buy 'a book worth looking at'—by which was meant one of the finer incunabula.

So highly did Morris value Walker's advice that in December 1889 he invited him to go into partnership with him as a printer. Walker refused to accept such an honor on the somewhat puzzling grounds that he 'had some sense of proportion.' Presumably he meant that he was so acutely aware of the disproportion between his own meagre talents and those of Morris that it would have been out of proportion for him to have become the formal partner of such a creative genius. But although he never became involved with the financial side of the venture, he was, according to Morris's secretary, S. C. Cockerell, virtually a partner in the Kelmscott Press from its first beginnings until its end. Morris's son-in-law was equally emphatic that Walker acted as, and virtually was, a partner in all but name, and took his full share from beginning to end of all the labors, cares, and anxieties involved. No important step was taken without his advice and approval, and such was the confidence placed in him that it was originally intended to have the books set up in Hammersmith and then printed in his office at Cliffords Inn.

Among the photographic enlargements supplied by Walker were some lines from Morris's copy of Aretino's *Historia Fiorentina*, printed at Venice in 1476 by Jacobus Rubeus in a type that closely resembled Jenson's. From this original Morris evolved the design of his Golden type (so named because it was to be used to compose the Kelmscott edition of *The Golden Legend*). After thoroughly absorbing the characteristics, virtues, and defects of the Rubeus fount by repeatedly drawing over the enlargements, Morris then drew the designs for his own type in the same large scale as Walker's enlargements. Next his drawings were photographically reduced by Walker to the scale in which the Golden type was to be cut. At this stage both Morris and Walker criticized them and brooded over them. Finally Morris worked over his drawings yet again until he was thoroughly satisfied with their design in every detail. He was at pains to explain that he did not make a servile copy of the Rubeus fount because he recognized that 'it is no longer tradition if it be servilely copied, without change, the token of life.' He set out to make his letters pure in form—'severe, without needless excrescences: solid without the thickening and thinning of the line, which is the essential fault of the ordinary modern type, and which makes it difficult to read; and not compressed laterally, as all later type has grown to be owing to commercial exigencies.' He was conscious too that his roman type, especially in the lowercase, tended rather more to the Gothic than did his fifteenth-century Venetian model.

The question of scale was all-important to him. Smoke proofs of the punches cut by Prince were given careful and prolonged examination. Morris used to go about with matchboxes in his pockets containing these smoke proofs, which he would pull out to scrutinize while talking with his family. In fact he had little reason to worry over the punchcutting, for he later declared that the punches for all his types had been cut with great intelligence and skill by Prince, who rendered his designs most satisfactorily onto steel. Matrices were struck by hand from these punches, but the subsequent mechanical casting of the type was one of the very few operations connected with the Kelmscott Press for which Morris allowed machinery to be used.

Typecasting was carried out in London by the firm of Sir Charles Reed & Son, managed at the time by Talbot Baines Reed, a prolific writer of school stories for boys, but also the author of a scholarly *History of the Old English Letter Foundries* which had been published in 1887. While Morris was designing his Golden type in 1890, Reed delivered a lecture at the Royal Society of Arts on 'Old and New Fashions in Typography' in which he developed his ideas on the relation between art and utility. Evidence that Morris appreciated the careful supervision which Reed gave to the casting of the Kelmscott types is provided by an album of photographs now in the St. Bride Printing Library with a penwritten note by Reed to explain that it contains 'enlarged photos of Early & Gothic type collected and presented to me by William Morris. 1891.' A further penciled note in what appears to be Reed's hand states that 'these types include the models upon which the founts were designed for use in the Kelmscott Press.' Reed's firm also cast the other two founts designed by Morris for his press. First came the Troy type, so named because its earliest use at the Kelmscott Press was to compose Caxton's translation of *The Recuyell of the Historyes of Troy*, published in 1892. The Troy type was cut in the latter half of 1891 from a design which Morris produced earlier in that year after a close study of fifteenth-century founts used by Peter Schoeffer at Mainz, by Gunther Zainer at Augsburg, and by Anton Koberger at Nuremberg. Troy departed even further from its sources of inspiration than the Golden type derived from the Rubeus fount. With his strong medieval tastes, Morris liked Troy better than Golden, and thought it just as readable. In Troy he avoided the early printers' liberal use of contractions, which he restricted to the ampersand (&) and to the few conventional ligatures such as ff, ffi, ffl which are commonly used by printers even to this day.

Readability is very largely a question of habit. The basic truth is that we read most easily the types that we read most frequently. In setting himself the task 'to redeem the Gothic character from the charge of unreadableness which is commonly brought against it,' Morris largely ignored the fact that Gothic letterforms were so infrequently read as continuous texts by his contemporaries that such letterforms, however beautifully and skilfully he interpreted them, were bound on account of their unfamiliarity to be considered less readable than roman type. His own satisfaction with his gothic type was that it provided a richer texture that was to him a pleasure in itself, and also made it a better foil for fine woodcut illustrations than any page of spindly roman type. For his final masterpiece of book production, he had the Troy design cut in a smaller size so that it could be used for composition in double column.

Having settled the design and manufacture of his types, the other basic materials needed for his press now came under his scrutiny. It will be convenient to discuss later his design of other typographical material such as woodcut initial letters, ornaments, borders, and illustrations. His first concern was to find the ideal paper, vellum, and ink.

Morris regarded it as a matter of course that the paper to be used at the Kelmscott Press would be handmade. He declared that 'it would be a very false economy to stint in the quality of the paper as to price,' and he therefore concentrated his mind upon the kind of handmade paper that he ought to use. He decided that it must be made entirely of linen (without any cotton such as was then commonly used when making paper by hand). In its finish, it was to be quite hard, through the use of plenty of sizing. And it was to be made on a mould so constructed that its wires would not produce too strongly ribbed an appearance in the sheets. He took as his model a sheet made by a Bolognese mill in 1473.

Accompanied by Walker he paid a visit to Batchelor's paper mill near Ashford in Kent. As soon as Walker introduced him to Joseph Batchelor, he recognized a fellow-enthusiast who was willing to act as his mentor, and who taught him the technique of papermaking. Morris actually made two sheets with his own hands, but as he found he could rely upon the mill to obtain what he wanted, he practiced no more and left it to Batchelor to carry out further experiments. The results were eminently satisfactory, so that in due course Morris obtained three papers from Batchelor and never used any others for the books printed at the Kelmscott Press. Each paper had a watermark designed by Morris, and in time the papers came to be known by the pictorial feature of the watermark. First deliveries were made on the following dates in the sizes shown parenthetically:

    12 February 1891   Flower (16×11 in.)
    22 April 1891   Flower (16×22 in.)

17 February 1893   Perch ($16\frac{3}{4} \times 22$ in.)

14 March 1895   Apple ($18\frac{1}{4} \times 12\frac{3}{4}$ in.)

The customary practice of producing 'large-paper' copies was one which Morris abandoned before setting up the Kelmscott Press, because the aesthetic effect of the type page was in his view ruined by surrounding it with an excessive amount of white space. For special copies he turned instead to printing upon vellum, thus continuing a tradition that goes back to Gutenberg's 42-line Bible. Vellum was a material he had used extensively for his calligraphical books. They had been written on the finest lambskin vellum obtained from Rome, where the Vatican's continuous demand for vellum had led to the traditions of its proper manufacture being more fully preserved there than anywhere else. Morris had used English lime-dressed vellum for a few special copies of *Love Is Enough* (1873), but he had found it disagreeable in surface and almost useless for fine work. Unfortunately the needs of the Vatican sometimes absorbed the entire local supply, and when the Kelmscott Press was set up, it was not possible to procure a sufficient quantity from Rome. Through a friend Morris was introduced to Henry Band of Brentford in Middlesex, who had experience of making binding-vellum, parchment, as well as heads for drums and banjos. Morris persuaded him to make some trials which were eventually completely successful. Band carefully chose skins of calves under six weeks old. He made Kelmscott vellum especially thin and gave it a special surface that was not faked with white lead. It was exceedingly expensive, but Morris regarded that as an unimportant detail provided the material met the requirements of the work to be done with it. When the growing needs of the Press exceeded Band's capabilities, the firm of William J. Turney & Co. in Worcestershire supplied vellum to Morris.

After the trouble he had experienced while making dyes, Morris knew that he would have some difficulties in obtaining an ink that would completely satisfy him; but the quest proved to be far more exasperating than he had ever expected. He was particularly irritated by 'those damned chemists' who he believed to have wrought infinite mischief in all matters of art. Often he spoke of making his own ink so as to be sure of its ingredients, but this intention was never carried out. Following a great many trials, two inks were obtained from manufacturers in England and in America. Neither was perfect because of their red or blue undertones. Yet both inkmakers reacted with the same take-it-or-leave-it attitude, and indicated that what was good enough for their other customers ought to be good enough for Morris.

Nothing but the best would ever satisfy Morris, and once again it was Walker who was able to indicate where the best material could be obtained. Many years earlier Walker had unsuccessfully tried every inkmaker in England before discovering that an excellent ink could be obtained from Jaenecke of Hanover. It was good in color, and if it showed any trace at all of weakening under months of daylight, it did not betray any unpleasing undertone. Its consistency was such that when a pinch of it was taken between finger and thumb, it could be stretched into a thread over an inch long. It never worked foul, nor clogged the type, nor spoiled the impression. There was, however, a snag. According to a letter written many years later by Walker to John Johnson, the University Printer at Oxford, Jaenecke's ink 'was tremendously stiff and very hard work for the pressman.' Walker further explained that its introduction at the Kelmscott Press produced something very near a strike, whereupon Morris, who was anxious to get on with the large number of books that were waiting to be printed, 'rather weakly gave in and used the best ink he could get in England. This turned out to be unsatisfactory, so Morris read the riot act and said if they wouldn't use the German ink he would close the Press. As his pressmen were very well paid they naturally did not want to lose their jobs and gave way. However, they took it out of Morris another way and the number of sheets turned out per day was extraordinarily small.'

Despite this incident, industrial relations at the Kelmscott Press were normally very good. Morris enjoyed talking and listening to his compositors, and has been described by an eyewitness as 'taking in every movement of their hands, and every detail of their tools, until he knew as much as they did of spacing, justification and all the rest.' He also spent hours with his pressmen, familiarizing himself with every peculiarity of their doings, such as the reason for damping paper in a given way, and to a given degree; but surprisingly he himself never pulled a sheet off his own presses, according to H. Halliday Sparling.

At the outset his sole employee was a retired master-printer, William Bowden, who was intended to act as both compositor and pressman. Only a few weeks later the amount of work to be done led to his being assisted by his daughter, Mrs. Pine, and later by his son. Following Bowden's retirement and the removal of the Press into larger premises, several new compositors were engaged. These included Thomas Binning from *The Commonweal*, a staunch trade unionist who was elected father of the chapel. His negotiations with the local union for membership of the entire Kelmscott Press staff forced the London Society of Compositors to accept Mrs. Pine as their first woman member.

As was formerly the custom in all printing offices, the staff was treated by their master to the traditional Wayzgoose feast every year from 1892 to 1895. For these occasions the compositors designed colorful menus for which they used their master's types and ornaments in arrangements that were a triumph of popular Victorian taste over their master's principles of typographical design. In fairness it must be said that their master's typographical interests were confined to book production, but in that field he had arrived at his own judgments 'from solid reasoning within himself,' as Moxon put it. Those judgments were expressed in his written note of 1894 on his aims in founding the Kelmscott Press, and at slightly greater length in a paper on 'The Ideal Book' which he read before the Bibliographical Society on 19 June 1893. In all the Kelmscott Press books the performance of the staff was rigorously directed in accordance with their master's principles: it is therefore appropriate to provide at this point a brief résumé of those principles.

Morris set out to print books in the hope of making some which, as he put it, 'would have a definite claim to beauty'; at the same time he wished them to be easy to read, not dazzling to the eye nor troublesome to the mind because of any eccentricity in the letterforms. He explained that it was only natural he should try to ornament his books suitably because he was a decorator by profession. But he always tried to keep in mind the need for making decoration part of the page of type. He laid it down that a book without ornament could look positively beautiful and not merely unugly, provided that it was architecturally good. By this he meant that the pages must be clear and easy to read, which they could not be unless the type was well designed, and unless the margins were in due proportion to the page of type. He had observed how many fifteenth-century books were beautiful purely by the force of their typography, even though many books of that period were lavishly supplied with added ornament.

Matters of spacing and position were considered by Morris to be of the greatest importance in producing beautiful books. If proper consideration was given to these matters, a book printed in a quite ordinary type would still look decent and pleasant to the eye, but disregard of these matters would spoil the effect of the best-designed type. He had been told that in the best medieval books and manuscripts, the amount of the four margins differed successively by twenty percent: the inner margin was always narrowest, the top twenty percent wider, the outer edge twenty percent wider than that, and the bottom widest of all. He attacked modern printers for systematically breaking this rule. In doing so they contradicted a fact to which Morris attached the greatest importance, namely, that the unit of a book is a pair of pages, not a single page.

Equal attention had to be paid to spacing between words and between lines of type. No more white was to appear between words than was needed to cut them off clearly from each other. Wider gaps between words tended to make a page illegible and ugly. He also opposed excessive space between lines, though he sometimes allowed a small amount of additional interlinear spacing with his gothic types, the Troy and the Chaucer.

He held firm views on what made a type legible. The letters had to be cast correctly upon their bodies, with very little white space between them. Excessive lateral compression was to be avoided in type design because it involved thinning the letters to a degree that made them illegible. Lowercase letters such as a, b, and d ought to be designed on something like a square to get good results. 'Letters should be designed by an artist not an engineer,' he wrote, and gave an example of the subtle difference between an i with a dot drawn by compasses, and the same feature delicately drawn in the shape of a diamond.

His own taste had been formed by a great admiration for the calligraphy of the Middle Ages and for fifteenth-century printed books. In roman type design his highest praise was for 'the generous and logical designs of the fifteenth century Venetian printers': Jenson's design struck him as being so clearly the best roman that had ever been made that it seemed to Morris a pity to make his own starting point at any period worse than the best. But once he had made his own roman 'as good as the best that has been,' he saw little scope for further development in that field. A certain amount of variety being desirable, he put in a word for some form of gothic letter, expressing a preference for the transitional forms used by early German printers. In gothic founts he observed that the lowercase was the strong side, whereas the strength of roman lay in its capitals, 'which is but natural, since the roman was originally an alphabet of capitals, and the lower case a deduction from them.'

He preferred to see a type-size not below twelve point used for an ordinary octavo, but he had a strong liking for larger formats. While admitting that some small books were tolerably comfortable, he considered that even the best of them were not as comfortable as a fairly big folio (by which he meant a format at least $12\frac{3}{4} \times 8$

inches). He complained that a small book seldom lay quiet, and that you either had to cramp your hand by holding it, or else had to put it on a table 'with a paraphernalia of matters to keep it down, a tablespoon on one side, a knife on another, and so on, which things always tumble off at a critical moment, and fidget you out of the repose which is absolutely necessary to reading. Whereas, a big folio lies quiet and majestic on the table, waiting kindly till you please to come to it, so that your mind is free to enjoy the literature which its beauty enshrines.'

The kind of literature enshrined in Kelmscott Press books was inevitably affected by the combination of such a patrician attitude to format, and of a typographical taste so strongly influenced by fifteenth-century models. Ronald Briggs has pertinently remarked that if Morris did not print the exploits of D'Artagnan, Jorrocks, or Huck Finn, all of which had given him many hours of delight, it was because they were not altogether in keeping with the stately format he preferred when making a beautiful book. In fact he was prevented from using a stately format as frequently as he would have liked. Analysis of the formats given to Kelmscott books reveals that Morris actually used 8vo or 16mo for over half the editions issued or planned before his death. To some degree this can be explained by a difficulty that he recognized when using double-column setting, for which folios are so well suited. Writing to his friend F. S. Ellis on 31 December 1890, he explained why one detail in a book to be set in his Golden type might disappoint Ellis: 'we cannot make a double-column page of it, the page will not be wide enough. For my part I don't regret it: double column seems to me chiefly fit for black letter.' He went on to draw Ellis's attention to the fact that Jenson did not print even his Pliny in double column, but stressed that in modern printing it was a case of *a fortiori* because we have no contractions and so few tied letters: 'we cannot break a word with the same frankness as they could: I mean we can't put whi on one side and ch on the other.' His only full folio, the great Chaucer, and his projected folio of *Froissart's Chronicles* were both set up in his gothic type, using the smaller size named Chaucer; and one reason why the Froissart was laid aside was the infeasibility of proceeding with two folios at the same time.

Analysis of the types used in Kelmscott Press books shows that practically equal use was made in them of roman and of gothic types. Further analysis reveals that by far the greater number of pages in all Kelmscott books were plain unornamented pages. Yet such is the power of the magnificent borders, decorated initials, and woodcut illustrations that it is these richly ornate pages which remain dominant in the public's mind and that reappear most frequently in textbook and catalogue reproductions of Kelmscott Press books. Because these pages are of such distinction, and because they reflect so well the individuality of their decorators, the attention given to them has been fully deserved. It follows that something must now be said about those who decorated the Kelmscott books.

By far the largest individual contribution was made by Morris himself: in little more than six years he designed a total of no less than 644 title pages, borders, decorative initials, and marginal ornaments. The astonishing ease with which he produced this work was observed by W. R. Lethaby: 'He would have two saucers, one of Indian ink, the other of Chinese white. Then, making the slightest indications of the main stems of the pattern he had in mind with pencil, he would begin at once his finished final ornament by covering a length of ground with one brush and painting the pattern with the other. . . . he seemed to have the idea that a harmonious piece of work needed to be the result of one flow of mind; like a bronze casting in which all kinds of patching and adding are blemishes. . . . The actual drawing with the brush was an agreeable sensation to him; the forms were led along and bent over and rounded at the edges with definite pleasure; they were *stroked* into place, as it were, with a sensation like that of smoothing a cat. . . . It was to express this sensuous pleasure that he used to say that all good designing was felt in the stomach.'

In style, S. C. Cockerell found that the borders and initials drawn by Morris for the first Kelmscott books were reminiscent of ornamentation in early fifteenth-century Italian manuscripts. However, Colin Franklin has noted a close resemblance between the border in the first Kelmscott book and the decoration upon the first page of an Appian printed by Ratdolt in 1477, while also noting that Morris drew his flowers in a more naturalistic style. A certain heaviness in his earliest decoration for Kelmscott books soon disappeared; backgrounds for initials were drawn in a lighter manner, with foliage on a larger scale, in order to correct the tendency of the earliest initials to appear too dark for the Golden type. To obtain the right balance he sometimes took the precaution of having a page of type proofed on drawing paper; he then drew his design in reverse directly upon the special proof, so as to concentrate his attention on the typographical effect, without being distracted by the wording.

All the designs for initials, ornaments, and illustrations in Kelmscott books were engraved on wood by W. H. Hooper, C. E. Keates, and W. Spielmeyer, except for a set of twenty-three illustrations by Walter Crane which A. Leverett engraved for

The Glittering Plain (1894), also a few early initials engraved by G. F. Campfield, and A. J. Gaskin's photo-engraved illustrations for the *Shepheardes Calender*. If initials or ornaments recurred, they were printed from electrotypes, but most of the title pages and initial words were printed directly from the wood. With very few exceptions the illustrations designed by Edward Burne-Jones, Walter Crane, and C. M. Gere were also printed from the wood. Only A. J. Gaskin's illustrations for Spenser's *Shepheardes Calender* (1896) and the facsimiles in *Some German Woodcuts of the Fifteenth Century* (1898) were printed from process blocks engraved by Walker's firm.

A highly successful method was devised for rendering Burne-Jones's illustrations onto wood. Most of the artist's original designs were drawn in pencil; they were redrawn in ink by R. Catterson-Smith (or in a few cases by C. Fairfax Murray) and were then returned to the artist for revision. Finally they were photographed onto wood for cutting by the hands of the craftsmen already named.

Such apparently tortuous methods to achieve typographical accuracy and harmony would have been fraught with danger had they not been controlled by a clear-minded and sensitive genius. Morris was admirably qualified to orchestrate the diverse talents of his chosen artists and craftsmen because the score remained so firmly fixed in his mind. As a voracious reader and a prolific writer, he was fundamentally concerned to produce books which would be a joy to read. Because he was also profusely talented as a decorator and endowed with great artistic talent, his own enjoyment would have been incomplete if all his books had been devoid of decoration or illustration. His ability to produce a personal and harmonious typographical style through his control of men and materials was solidly based; his detailed knowledge of early manuscripts and printed books provided a well from which he drew extensively, but to which he added by a long period of creative experiment in calligraphy, book design, and type design. His practical experience as a craftsman and manufacturer had given him experience in how to obtain the finest colors and materials. This had led to the establishment of high standards from which he never departed during his typographical adventure. Furthermore, his life was enriched by deeply held convictions as to the purpose of life and the way in which it ought to be conducted. His friendship at Oxford with Edward Burne-Jones who, like Morris, abandoned studies for the Church in order to devote himself to art, led to a lifelong association which culminated in the finest of all the Kelmscott books.

Burne-Jones referred to the Kelmscott Chaucer as 'a pocket cathedral—it is so full of design.' Much of its design was from the hand of Morris, who complemented the set of eighty-seven Burne-Jones illustrations by designing a woodcut title, fourteen large borders, eighteen different frames round the illustrations, and twenty-six large initial letters. He also completed a design for its binding in full white pigskin over oak boards which Cobden-Sanderson executed from 1896 on forty-eight copies, a task only completed some years after the death of Morris. Cobden-Sanderson had set up his Doves Bindery in 1893 close to Kelmscott House, attracted partly by the opportunity to repair and bind books in Morris's library, and also in the hope of working on some Kelmscott Press bindings, which were normally carried out by J. & J. Leighton, either in a plain half holland style or in simple vellum bindings.

Morris had the highest regard for what he termed the 'magnificent and inimitable' illustrations made for him by Burne-Jones. These gave Morris particular satisfaction because they formed not only a series of most beautiful and imaginative pictures, but also made 'the most harmonious decoration possible to the printed book.' This harmony was to a large degree achieved by the sensitivity with which Morris contrived his superb borders as frames for his friend's illustrations; much also stemmed from the technical skill with which the illustrations were rendered upon wood, and the care with which they were so snugly fitted into their border-frames. Of the thirteen books illustrated by Burne-Jones for the Kelmscott Press, the Chaucer is the largest in size and in range, and it is deservedly the most highly praised of all Kelmscott books. The text was printed in black and red; the Chaucer type was set in double column, with headings to the longer poems in the Troy type. Thirteen copies were printed on vellum, 425 on paper. No modern private press book has been so frequently reproduced, and in recent years there have been two facsimile editions.

Walter Crane only drew one set of illustrations for a Kelmscott book; engraved on wood by A. Leverett, they appeared in the 1894 edition of *The Story of the Glittering Plain*. As engravings they appear unduly heavy and lacking in sharpness—defects which may explain why Leverett was not engaged again, and also why Crane himself made no mention of them in his own treatise on *The Decorative Illustration of Books* (1896) in which he merely included one reduced reproduction from the set of twenty-three. Crane was noncommittal about Burne-Jones's work for the Kelmscott Press, but had something to say about three other Kelmscott illustrators all from the Birmingham School: C. M. Gere, E. H. New, and A. J. Gaskin. Though Crane did not mention Gaskin's illustrations for *The Shepheardes Calender* which only appeared in the year when Crane's treatise was published, he praised Gaskin's earlier work. A recent critic, Colin Franklin, has extolled Gaskin for producing 'the happiest of all Kelmscott illustrations, bringing a strictness of form and composition none of the others provided.'

Seven different formats were used for Kelmscott editions, but within this diversity of sizes were some uniform sets or series, or groups linked by a common theme. First came a group of five Caxton editions issued in the years 1892–93: three were published in large quarto by Bernard Quaritch; the next in small quarto was sold by Reeves & Turner while the fifth volume, again in large quarto, was published at the Kelmscott Press by Morris. Together they form a handsome tribute to England's first printer-publisher. Three translations made by Morris from the French appeared in uniform 16mo format, but his sixteen original texts were printed at Kelmscott in a variety of formats. Two sets of particular attractiveness were printed in octavo and edited by F. S. Ellis. The first was a series of English poets printed in Golden type. Later came a trio of reprints from the Camden Society's volume of Thornton romances: *Syr Perecyvelle of Gales* (1895), *Sire Degrevaunt* (1896), and *Syr Ysambrace* (1897). All three were printed in red and black with Chaucer type, and each was decorated with a woodcut designed by Burne-Jones. Color other than red was rarely used in Kelmscott books; red for initials appeared only once, in the volume of Wilfrid Scawen Blunt's poetry, and then it was evidently done at the author's insistence. 'It looks very gay with its red letters,' Morris wrote to his daughter Jenny, 'but I think I prefer my own style of printing.' Initials in two colors were mentioned by Morris in another letter written in January 1893 concerning his projected edition of *The Tale of King Florus*; but by the time it was published in December of that year, the idea had been abandoned, though some of Morris's experimental designs for two-color initials have survived. Perhaps his reaction to Blunt's volume the previous year had turned him temporarily against such abundant use of color in his printing. But for the fact that Blunt was a fellow-poet, Morris might never have been prepared to make any concession in his style of printing, for he had made it absolutely clear in his publishing agreement signed with Bernard Quaritch in 1890 that he himself was to have sole and absolute control over typographical matters.

Exercise of sole and absolute control gave the Kelmscott books such powerful individuality, as well as such compelling coherence of design, that they have exerted an influence upon book typography in Europe and in the United States which has gone beyond that of any other designer or press. On both continents the impact of Morris on book design came about in three ways: through imitation, through a desire to make books in accordance with his principles, and through his effect upon the course of type design.

Without any original there can be no imitation, as the Grossmiths tersely observed. Numerous imitations made of Morris's typographical material—his decorations and his types—are nevertheless unwelcome as tributes to his originality. Imitations are essentially servile copies, and these Morris had rigorously opposed because they were inevitably made 'without change, the token of life.' Admittedly there are occasions when it is hard to draw the line between a design that is strongly influenced by another and one that at first glance appears to be an imitation. Nevertheless, it would be unrewarding to dwell upon either imitations or near-imitations: only a few will be mentioned while tracing the effect through Western Europe and North America of the typographical example set by William Morris.

The Kelmscott Press was kept going for about eighteen months after the founder's death in 1896. With Walker's assistance, the two trustees F. S. Ellis and S. C. Cockerell finished off the work in hand and produced the more important works for which Morris had made preparations. Once that had been accomplished it was wisely decided to close the Press; but within a few years, Walker, Cobden-Sanderson, and Cockerell were actively continuing its traditions through their association with other private presses or trade publishers, and through their influence as advisers and teachers.

Walker went into partnership with Cobden-Sanderson in 1899 to start the Doves Press. For their type, the Rubeus photograph, previously used by Morris while designing his Golden type, was retouched by Walker's draughtsman, Percy Tiffin, but this time with the intention of producing a closer, less gothic version of the original Venetian type. Once again the services of Edward Prince were secured, and punches were cut by him in faithful conformity to Tiffin's cleaned-up drawings. Nevertheless, the punchcutter's skill helped to breathe into the type a degree of liveliness that it might otherwise have lacked, thereby proving the truth of the Gothic lesson taught by Morris: that a vital tradition can take inspiration and encouragement from the past, and that the best achievements of the past can be created anew—which is a very different matter from servile copying or imitation.

Superficially the typography of Doves books was in marked contrast with those from the Kelmscott Press: no gothic type, no ornament, but a reliance on 'pure' typography relieved only by hand-drawn lettering. Yet there could still be seen the same Kelmscott practice of closely set type of Venetian style, carefully proportioned margins, fine presswork, and excellent materials.

While Walker was in partnership with Cockerell in 1901, his firm undertook the making of a type for the Ashendene Press which C. H. St. John Hornby had started in a modest way in 1894. An afternoon with Morris at the Kelmscott Press in 1895 made a lasting impression on Hornby who found himself in full agreement with the essential canons of good printing upon which Morris had constantly insisted. Hornby believed that in the forty years after their meeting, the spirit which infused the work of Morris had done more than anything else to influence the printing craft not only in England but throughout the world.

A few more examples must be given of his influence in England. It extended to other private presses such as his friend C. R. Ashbee's Essex House Press (1898–1909) which employed some of the former Kelmscott staff, but regrettably also used two hideous types designed by Ashbee and cut by Prince; more happily it influenced the publishing firm of J. M. Dent which commissioned the Lanston Monotype Company in England to produce a Venetian-style type named Veronese, completed in 1911, and which also published the Everyman Library with decorations drawn in the Kelmscott manner by Reginald Knowles. These few examples cannot adequately convey the extensiveness of Morris's influence in England, simply because his example and his writings had as great an influence as that exerted by those who had known him and worked with him.

In turning to his influence abroad, more must be said about the work of his punchcutter, Edward Prince. As well as cutting types for the English private presses already named, Prince cut types for the Vale and Eragny presses as well as types for two London publishers—Florence for Chatto & Windus and Riccardi for the Medici Society. Furthermore, Prince cut types for the Cranach Press in Germany and for the Zilverdistel Press in Holland. The commission for the Cranach Press came through Walker, whose advice had been sought by Count Harry Kessler, both for his private press in Weimar, and for the Insel Verlag with which he was closely connected. One of the finest Cranach Press editions was dedicated to Walker by Kessler who described him as 'the master of the art of the book, the inspirer and friend of William Morris.'

Morris's typographical influence reached the Low Countries through Antwerp where Emmanuel de Bom, cofounder of the literary magazine *Van Nu en Straks*, gave a lecture in 1904 at the Plantin-Moretus Museum on the subject of Morris's influence on the book, to coincide with the opening of an exhibition on the same theme. The Belgian designer Henry van de Velde was inspired by Morris's view of the book as an architectonic unit; the Morris influence can also be seen in his manner of decorating books, but Van de Velde drew less heavily on the past for his inspiration. Similarly the Dutch artist, type designer, and book designer S. H. de Roos, whose first book designed in 1903 was an edition of some essays by Morris translated into Dutch, venerated the great Englishman's attitude and principles but evolved a typographical style that owed more to Morris's precepts than to his Kelmscott examples.

In Scandinavia the impact of Morris was seen to a marked degree in the work of the Swedish printer Waldemar Zachrisson, whose printing showed how a commercial firm could put Morris's principles to practical use and also how his style could be adapted to suit Swedish taste. In Denmark the work of Walker and Cobden-Sanderson had a particularly strong impact on C. Volmer Nordlunde, a master printer who even found the time to write, design, and print an excellent pair of monographs on his two English heroes. Only in France did Morris seem to make no impact upon typographical practice.

Nowhere in the world did Kelmscott books make a faster or deeper impression during the 1890s than in the United States. Along its eastern seaboard and also in Chicago, the time was ripe for typographical innovation among printers, publishers, and book collectors. A number of them banded together in clubs such as the Grolier in New York, the Philobiblon in Philadelphia, and the Club of Odd Volumes in Boston. Their tastes were whetted by publications such as *The Knight Errant* (Boston, 1892–93) with its glowing reports on Kelmscott books, and by a spate of articles in other longer-lived periodicals. Two early examples of Kelmscott influence upon the design of American books can be found in F. Hopkinson Smith, *A Day at Laguerre's* (Houghton Mifflin, 1892) and D. G. Rossetti's *The House of Life* (Copeland and Day, 1893). In Chicago, Kelmscott-printed copies of Rossetti's *Hand and Soul* (1895) were published by Way and Williams, after a discussion in London between Morris and W. Irving Way, but this was the only occasion when Kelmscott books were supplied with an American imprint.

One American enthusiast who went the whole way from early imitation of Kelmscott books to later and more personal interpretation of their underlying principles was the Boston scholar-printer D. B. Updike. In August 1893 he wrote to ask

Morris if he could obtain a supply of type, or copy it, so that it could be used in a devotional book which he sincerely desired to make 'the most perfect piece of printing that is possible.' Apparently Morris left his request unanswered, but a copy of the Golden type was nevertheless used for a circular published by Updike in October 1894.

The copy was to have been his exclusive property: he had arranged for its manufacture by J. W. Phinney at the Dickinson Type Foundry in Boston, but after learning that a quantity of it had been supplied to another printing house before the design had received his final approval, Updike decided to use it only for his 1894 circular. Two years later when his *Altar Book* was published it was set in the Jenson-style Merrymount type that he commissioned from the architect Bertram Goodhue, who also drew Kelmscott-style initials and borders for the book. Updike later regretted the derivative work he produced during this period, but it attracted favorable comment at the time from some of his contemporaries. One American publisher, J. M. Bowles, went so far as to send trial pages to Morris in order to obtain his opinion on a deliberately Kelmscott-style volume that he published in 1895. For this work, the title page, headbands, and initials were drawn by Bruce Rogers, to whom Bowles's collection of Kelmscott books had come as a revelation.

Possessed as he was of such exceptional creative powers, Rogers never fell into the trap of copying the work of Morris. To borrow a striking phrase from Beatrice Warde, 'We owe an immensely greater debt to Mr. Rogers for having managed to steal the Divine Fire which glowed in the Kelmscott Press books, and somehow to be the first to bring it down to earth. . . .' But is it possible to define the nature of that fire, or to explain how it can be brought down to earth? In essence I believe it to have been the ability of Morris to inspire others, and this stemmed from his passionate love of craftsmanship and fine materials. Through becoming a public figure, he captured the imagination of many who never met him—by his use of words, his love of books, and his burning desire to improve the world.

The Kelmscott example showed what had to be done in order to improve typographical quality. Some like Bruce Rogers went back to a similar Jenson model for type, but by adapting its design for mechanical composition brought it within the reach of trade printers. Others like Francis Meynell understood the need for good paper, and for adding suitable decoration and illustration, but by meeting these needs in large editions and at modest prices brought the Kelmscott example to a wider readership. There were others too like Jan Tschichold who were as scrupulous in their concern for evenly spaced composition and for carefully regulated margins, but through obtaining these features from paperback printers brought the Kelmscott example to the mass market.

Thanks to the achievements of Morris at his Kelmscott Press, the role of a typographer came to be understood and acknowledged by printers and publishers on both sides of the Atlantic. Bruce Rogers was but one in a long line of typographers in various countries, and with widely differing outlooks, who placed on record their debt to the inspiring quality of his work. That was its Divine Fire.

CATALOGUE

The Library of William Morris

THE MANUSCRIPTS AND ILLUSTRATED INCUNABULA EXHIBITED are all from the collection of the Pierpont Morgan Library, repository of the largest and most valuable single block of volumes from Morris's library. The manuscripts are referred to by their Morgan Library manuscript (M.-) numbers, and the incunabula by their accession numbers, to which are added references to F. R. Goff, *Incunabula in American Libraries: A Third Census*. The woodcuts in the incunabula have been counted, numerations of the form '92/97 woodcuts' meaning that 92 distinct cuts appear, some of which are repeated for a total of 97 occurrences. Morris's library was famous for its German illustrated books, but an attempt has been made here to show also the importance of his early French books. The manuscripts and printed books below are described more fully in the 1906–07 catalogue of Pierpont Morgan's library (no. 13), and the manuscripts are also listed in Seymour De Ricci, *Census of Medieval and Renaissance Manuscripts in the United States and Canada*.

Documentation

1

CATALOGUE OF MORRIS'S LIBRARY, c. 1876
18 leaves, 9 × 6⅛ in. [Paul Mellon]

The catalogue is partly in the hand of William Morris and partly in that of (probably) May Morris. Though undated, its most recent entry is for a publication dated 1875. Thirteen manuscripts (seven of them Morris's own calligraphic work) and 291 printed books are entered, the last 54 of which are Icelandic.

2

CATALOGUE OF MORRIS'S LIBRARY, 1890–91.
93 leaves, 12¾ × 8 in. [Bernard H. Breslauer]

The first entries, in pencil, are in the hand of Jenny Morris; the entries on fo. 72 and after, in ink, are in Morris's hand, and probably represent current acquisitions. In Jenny Morris's section rare books and modern books are intermixed. Morris went back through this section, marking incunabula with an X, early sixteenth-century continental books with XX, and later continental and sixteenth-seventeenth-century English books with XXX.
PLATE I

3

CALLIGRAPHIC CATALOGUE OF
MORRIS'S LIBRARY, c. 1890
8+12 leaves, 9¼ × 7½ in.
[Bernard H. Breslauer]

This incomplete catalogue is written on a gathering of eight leaves, in Morris's autograph. Four ornamented initials have been finished in color; the remainder are roughly sketched in pencil. Twelve manuscripts (six of them Morris's own calligraphic work) and fifteen incunabula are entered. Sydney Cockerell had the catalogue bound (by Katharine Adams) together with a gathering of twelve leaves, three pages of which contain Morris's incomplete calligraphic transcription of *The Story of Haldor*. This is on the same stock of Whatman paper as the library catalogue, and probably was done at about the same time.

PLATE II

4

COCKERELL'S CATALOGUING
NOTES
10 leaves, varying sizes.
[PML: gift of John M. Crawford, Jr.]

A small group of rough notes made by Cockerell while cataloguing Morris's books has survived, some with added annotations by Morris. Several (including that illustrated) are attempted schemas of the Augsburg imprints. Another, the census of woodcuts in the Nuremberg Chronicle, was reprinted in *Some German Woodcuts*, 1897 (no. 11).

PLATE III

5

SALE CATALOGUES MARKED BY
MORRIS
Six catalogues bound together. [Grolier Club]

Sydney Cockerell preserved Morris's annotated copies of six London auction catalogues: for the sales of Edward Hailstone (Sotheby's, 23 April 1891), E. H. Lawrence (Sotheby's, 9 May 1892), T. and W. Bateman (Sotheby's, 25 May 1893), Fountaine (Christie's, 6 July 1894), Howell Wills (Sotheby's 11 July 1894), and William Stuart (Christie's, 6 March 1895). On the cover of one Morris also scribbled architectural notes which may have been made at an S.P.A.B. meeting.

6

MORRIS'S DESCRIPTIONS OF HIS
BOOKS

Morris began writing out entries for his projected printed catalogue in May 1894. He completed a number of short descriptions of his woodcut-illustrated incunabula, but only four descriptions of his manuscripts are known.

A   MANUSCRIPT NOTES. 1 leaf, 6⅜ × 8 in.
    [PML: gift of John M. Crawford, Jr.]

The description illustrated is for the Latin-German *Speculum humanae salvationis* printed at the monastery of SS. Ulrich and Afra in the types of Günther Zainer. It was separated from Morris's copy of the book (see no. 26) after the 1898 sale.

PLATE III

B   TRIAL PAGES FOR THE PRINTED
    CATALOGUE [British Library]

At least two trial pages of the planned catalogue were set up and printed at the Kelmscott Press, and preserved by Sydney Cockerell. One entry describes the French Psalter now at the Morgan Library, M.101 (no. 23); the other describes the Augsburg *Speculum humanae salvationis* (no. 26). This latter entry shows particularly clearly how Cockerell's detailed bibliographical notes were intended to be joined to Morris's more general descriptions.

7

COCKERELL'S CATALOGUE OF THE LIBRARY
(a) Ledger book, 9×6⅝ in. (b) 12 leaves, 9×7¼ in. [PML: gift of John M. Crawford, Jr.]

Sydney Cockerell never entirely completed his cataloguing of Morris's library. In a ledger book he entered, alphabetically by city, 'Early books with woodcuts.' On loose ledger leaves he separately listed 'Early printed books,' i.e., without woodcuts. In several books *sine nota* he added in red ink attributions of printers made by Robert Proctor. These catalogues may have been made after Morris's death. In his diary for 26 April 1897, the day before the library was packed up by Pickering and Chatto, Cockerell records being 'Very busy all day copying my catalogue notes so as to have some record of the books.'

8

THE LIBRARY, KELMSCOTT HOUSE

This photograph was taken in the second half of 1896. The manuscript on the reading stand in the foreground is the Windmill Psalter (no. 25); to the left, closed on the table, is Günther Zainer's edition of the *Spiegel des menschlichen Lebens* (no. 28). The upper shelves of one case are mostly filled with vellum-bound Kelmscott Press volumes. In the bottom shelf of that case, partly obscured by the table, the tallest volume is part of a large folio French Bible (no. 22), newly bound with white pigskin spine by Douglas Cockerell; its two companion volumes, in older plain vellum, are next to it.

PLATE IV

9

F. S. ELLIS'S INVENTORY OF THE LIBRARY
53 leaves, 13×8¼ in. [Sanford L. Berger]

This priced inventory was made by Ellis and his son the week after Morris's death. Though Cockerell's note says prices 'were as near as possible those paid by Morris,' the bargain acquisitions were readjusted to fair market value, and in a few instances the assigned prices are lower than Morris actually paid. The running totals are not entirely accurate, but the MSS were appraised at about £12,000, the printed books (not including modern works) at about £7300. There are additional annotations by Cockerell and Fairfax Murray.

PLATE V

10

FAIRFAX MURRAY'S ATTEMPT TO BUY THE LIBRARY
9 documents including 5 letters, 16 December 1896 – 26 January 1897. [PML]

Fairfax Murray proposed in October 1896 to buy Morris's library *en bloc*; formal negotiations were made in December, but fell through owing to Murray's inability to put up full security for the selling price, £20,000. The Morgan Library owns several draft agreements for this purchase and correspondence among the principals—Murray, Ellis, and Cockerell—showing why the agreement broke down.

11

SOME GERMAN WOODCUTS OF THE FIFTEENTH CENTURY
Kelmscott Press, 1897.
[PML: gift of John M. Crawford, Jr.]

Morris did not live to complete the long-planned printed catalogue of his library. Some idea of what was intended can be gotten from the posthumous *Some German Woodcuts*, put together by Cockerell with the assistance of Robert Proctor. It includes twenty-nine electrotype reproductions chosen by Morris to illustrate his catalogue, and a check-list of his most important woodcut-illustrated incunabula. This copy was presented by Cockerell to W. R. Lethaby.

12

THE 1898 SALE OF MORRIS'S BOOKS
Sotheby, Wilkinson & Hodge, 5–10 December 1898. [Grolier Club]

After purchasing Morris's library in April 1897, Richard Bennett kept only a selection of the books. The majority, including some items of great interest and value, were auctioned at Sotheby's the following year. The sale, in 1215 lots, realized a total of £10,992 11s. The highest price, £350, was brought by the fourteenth-century Sherbrooke Missal, now in the National Library of Wales.

13

MORRIS'S BOOKS AT THE PIERPONT MORGAN LIBRARY
Catalogue of Manuscripts and Early Printed Books from the Libraries of William Morris, Richard Bennett, Bertram, fourth Earl of Ashburnham, and Other Sources Now Forming Portion of the Library of J. Pierpont Morgan. 4 volumes. London, Chiswick Press, 1906–07. [PML]

Pierpont Morgan bought the whole of Richard Bennett's library in the summer of 1902. He then commissioned this catalogue, probably the most elaborate ever made of a private collection. The bulk of the volumes are from Bennett's library, and include twenty-nine medieval manuscripts and 239 early printed books once Morris's. On exhibit is one of the five copies of the catalogue printed on vellum.

Manuscripts

14

PSALTER
London, end of the thirteenth century. $8\frac{5}{8} \times 6\frac{3}{8}$ in. [M.100]

This Psalter is one of a small group of English manuscripts whose decoration is confined to illuminated initials and dynamic border ornament, combined with coats of arms. It was probably made for William, first Baron de Vescy, who died in 1297. The manuscript was bought by Morris from Quaritch probably in 1892 or 1893. In his description made for his projected catalogue Morris wrote of it, 'Though this book is without figure-work the extraordinary beauty and invention of the ornament make it most interesting; and the said ornament is thoroughly characteristic of English work: the bold folded-over leaf in the great B on the first page of the text is an undeniable token of an English hand. The great thickness of the black boundary lines is worth noting, and no doubt is the main element in producing the effect of the color, which is unusual. . . .'

PLATE VI

15

PSALTER
Northern France, late thirteenth century. $9 \times 6\frac{1}{4}$ in. [M.98]

It appears that Morris bought this Psalter privately, probably before 1894, from Charles Butler, who in any case owned the manuscript in 1888. The Kalendar is lacking, making localization very difficult. Morris completed a description of it for his catalogue: 'This book has a complete and satisfactory scheme of ornament, which is nowhere departed from, and the colour of which is thoroughly harmonious. Many of the dragon-scrolls end in daintily painted little heads, drawn with much expression and sense of fun; and the hair of them beautifully designed, and drawn very firmly. The figure-work in the eight historiated letters is everywhere quite up to the average of its date, but on the first page in the *Beatus* and the symbols of the Evangelists goes a good deal beyond that. Altogether an admirable specimen of the work of the later thirteenth century.'

PLATE VII

16

BOOK OF HOURS, Use of Verdun
France (Verdun?) c. 1375. $5\frac{1}{8} \times 3\frac{7}{8}$ in. [M.90]

This small Book of Hours was bought by Morris at Quaritch's on 22 April 1895, probably for £84. The Kalendar is in French, as are additional prayers at the end. The delicately colored miniatures, by a follower of the Maître aux Bouqueteaux, are on separate pieces of vellum that have been pasted in the manuscript.

PLATE VIII

17

TWO MINIATURES ON A BIFOLIUM
Augsburg, Johannes Bämler, 1457. $11\frac{3}{4} \times 16\frac{3}{4}$ in. [M.45]

Morris acquired only a few German manuscripts. He bought these conjoint miniatures from Quaritch in the spring of 1895, because of their association interest with Augsburg printing. The miniatures represent St. Leonard freeing prisoners, and a Crucifixion. Sydney Cockerell suggested the sheet was the opening of a book for a confraternity of St. Leonard. The crucifixion miniature is signed 'Artifex Johannes Bemler 1457.' Bämler was a scribe, rubricator, and illuminator who in the 1470s made a smooth transition to typography. He printed many interesting vernacular texts, often with woodcut illustrations. Morris owned six illustrated books from his press.

PLATE IX

18

PSALTER
Possibly made at Lesnes Abbey, Kent, first quarter of the thirteenth century. $12\frac{1}{2} \times 9\frac{1}{4}$ in. [M.43]

This Psalter may have been made for Lucie and Roger Huntingfield, whose obits occur in its calendar and after whom the manuscript receives its name, the Huntingfield Psalter. Its cycle of illustration is one of the most interesting and lavish produced in England at that time. Morris acquired it from Quaritch in May 1895 for £800, almost three years after first seeing it, and following three distinct decisions not to buy it.

PLATE X

19

MISSAL, Use of Sarum
England, third decade of the thirteenth century. $16\frac{5}{8} \times 12$ in. [M.107]

This splendid codex, known as the Tiptoft Missal, was commissioned by John Clavering and his wife, Hawyse Tiptoft. The arms of both families occur regularly in the manuscript's uncommonly lavish borders. M. R. James wrote of it, 'it would be difficult to find a more magnificent specimen of the English art of the period than this book affords.' The Tiptoft Missal was acquired by Morris in May 1895 at Sotheby's, apparently by private treaty, at a price of £900.

PLATE XI

20

ROMAN DE LA ROSE and
TESTAMENT DE JEAN DE MEUNG
France, c. 1380. 8 × 5½ in. [M.132]

The text is in the Burgundian dialect. The seventy-one grisaille miniatures are related to the style of the Maître aux Bouqueteaux. Morris acquired the manuscript, for £400, from Quaritch in June 1895. There are some indications that he vacillated over its purchase. He signed his name on a flyleaf with the date 20 June 1895. Four days later Cockerell returned a *Roman de la Rose* manuscript to Quaritch, but then on 4 July, at Gatti's, 'WM had the new MS. Romance of the Rose there.'

PLATE XII

21

PSALTER
Flanders, probably Ghent or Bruges, third quarter of the thirteenth century. 9½ × 6½ in. [M.106]

Bought on 2 October 1895, when Cockerell recorded in his diary, 'WM came up from Kelmscott to see an MS. Psalter at Ellis & Elvey's which he bought for £325—Flemish with fine pictures—he went back in the evening.' The Psalter may have been made for the convent of the Poor Clares in Bruges. A number of closely similar manuscripts are known from the same Flemish workshop.

PLATE XIII

22

BIBLE IN LATIN
3 vol. France, c. 1260. 20 × 14½, 18¾ × 13, 18½ × 13 in. [M.109–111]

Morris owned several folio and large-folio French Bibles of this period. This one was bought from Leighton in December 1895 for £650. Despite its large format, it was retained by Richard Bennett at least until 1900. He then seems to have got rid of it, for when Pierpont Morgan bought Bennett's library two years later it was not in the collection, and hence is not described in Morgan's 1906–07 Catalogue. A deleted entry in an early Morgan Library accession book seems to indicate that Morgan acquired it from Quaritch at an unknown date. Three more manuscripts from Morris's collection have come to the Morgan Library since then: an Onosander (M.449), a large folio Josephus (M.533–34), and most recently a thirteenth-century Bible, Kings only (M.968), the gift of John M. Crawford, Jr.

PLATE XIV

23

PSALTER
Northern France, probably Beauvais, c. 1250. 8⅛ × 5⅝ in. [M.101]

This Psalter was bought in December 1895 from Ellis & Elvey for £375, 'an excessive price' according to Cockerell at the time.

Following its Kalendar are ten leaves of full-page paintings, four subjects to a page, illustrating the life of Christ. Morris completed a description of this manuscript: '. . . These ten leaves of designs, clearly done by a different hand from the other figure-work, are of the best French work, at once elegant and serious. The historiated letters are clear and bright in colour, which tends towards the English in character, the figures large in scale for their spaces, and quite firmly drawn; there is a particular charm about these letters, which take up a good space on the page. . . . The writing throughout the book is big, and bold, and as good as can be.'

PLATE XV

24

BESTIARY AND OTHER TEXTS
England, perhaps Lincoln, before 1187. 8½ × 6⅛ in. [M.81]

This beautiful Bestiary was acquired by Morris in April 1896 from Jacques Rosenthal of Munich, for £900. It had been in the London sale rooms in 1889, when the Prussian government auctioned off a portion of the Hamilton Palace manuscripts, which they had bought *en bloc* in 1882. The Bestiary was one of a group of manuscripts donated to Radford (now Worksop) Priory in 1187 by Philip, Canon of Lincoln. Philip's anathema against anyone moving the book beyond the precincts of the priory is written on the first leaf. This Bestiary has a close textual relationship, first discovered by Sydney Cockerell, with one in the British Library, Royal MS. 12 C XIX.

PLATE XVI

25

PSALTER
England, perhaps East Anglia, late thirteenth century. 12¾ × 8¾ in. [M.102]

This magnificent Psalter may be appropriately thought of as the crowning jewel of Morris's library, for it was his last and finest acquisition. He bought it privately in July 1896 from Lord Aldenham, for £1000. The month before, at a Society of Antiquaries exhibition, Morris had recognized it as the source of a fragment of four Psalter leaves already in his possession. Because of the unusual representation of a windmill on the second folio, Morris named this the Windmill Psalter. It has been surmised but never proven that the windmill was a badge of original ownership.

PLATE XVII

Incunabula

26

[AUGSBURG] SPECULUM HUMANAE SALVATIONIS
[Latin-German]. Monastery of SS. Ulrich and Afra, not after 1473. f°. 176/192 woodcuts. [PML 129, Goff s-670]

The anonymous *Speculum humanae salvationis* has an interesting tradition of illustration embracing manuscripts, block books, and printed books, on which see also nos. 33 and 44. This edition uses Günther Zainer's types and initials, but a 1473-dated note in one copy of the book says it was printed in the Monastery of SS. Ulrich and Afra, which later did other printing with borrowed types. The work was also edited and translated in the monastery, the German version being for 'nuns and other religious persons who do not understand Latin.' Morris's description of the book (no.

6) says the cuts 'are rude, but have much character though they are not so decorative as some others of the Augsburg & Ulm books.' The present copy was sold to Pierpont Morgan as being Morris's, but there is a difficulty: Morris's copy, disposed of by Bennett in 1898, had (in Morris's words) 'systematic & good' coloring of the woodcuts, whereas the Morgan copy has only occasional light coloring.

PLATE XVIII

27

[AUGSBURG] JACOBUS DE VORAGINE. LEGENDA AUREA

Günther Zainer, c. 1475. f°. 140/162 woodcuts. [PML 134, Goff J-84]

The woodcuts in this Golden Legend were first used in Zainer's 1471–72 German edition of the same text, which was the first illustrated book printed in Augsburg. This reprint was the only illustrated Latin edition of the Golden Legend printed in the fifteenth century. In his description of the book Morris wrote, 'The cuts in this book are done by an early artist of Zeiner's and have the look of being designed by an illuminator; they look like the work of Sorg's Speculum: many of them are very pretty, and they are decidedly ornamental in character, though lacking in the freedom of the artist of the [Spiegel des] menschlichen Lebens. The type though we have called it semi Gothic might rather be looked on as a slight alteration towards Gothic of his first Roman type.'

PLATES XIX, XXXIX, XL

28

[AUGSBURG] RODERICUS ZAMORENSIS. SPECULUM VITAE HUMANAE

[German]. Günther Zainer, c. 1475. f°. 55/56 woodcuts, one full-page. [PML 139, Goff R-231]

Acquired by Morris 1 August 1890. In his long description of the volume, inserted in this copy, he wrote, 'Its matter (the contrast of the good & the bad side of all degrees and careers) makes a good subject for the illustrator, who has made the most of the opportunity; and indeed I should call this the best of the Augsburg picture books.... These cuts together with the beautiful "Blooming-letters", and Gunther Zeiner's bold and handsome second Gothic type make up as good ornamented pages as are to be found in all typography.' The same woodblocks were reused in Augsburg (1479), then in Lyons (1482), and finally in Zaragoza (1491).

PLATES XX, XXI

29

[AUGSBURG] BOCCACCIO. DECAMERONE

[German] Anton Sorg, 18 October 1490. f°. 87 woodcuts, one full-page. [PML 162, Goff B-731]

This was the second edition of the *Decameron* in German, translated by Heinrich Schlüsselfelder of Nuremberg. The first edition, Ulm, c. 1473, was not illustrated. Anton Sorg was one of the most prolific of fifteenth-century printers, a high proportion of his output being of vernacular works with woodcut illustrations.

PLATE XXII

30

[ULM] BOCCACCIO. DE CLARIS MULIERIBUS

Johann Zainer, 1473. f°. 79/81 woodcuts. [PML 194, Goff B-716]

This Boccaccio was probably the first illustrated book printed in Ulm, standing thereby at the head of a great tradition. Zainer used the same woodcuts in a German edition, whose dedication is dated 15 August 1473, and in a very curious picture-book edition, in which the cuts were printed two to a page with brief captions in the place of text. Morris wrote that the Ulm Boccaccio 'is a very old friend of mine, and perhaps the first book that gave me a clear insight into the essential qualities of the mediaeval design of the period. . . . the composition [of the woodcuts] is good everywhere, the drapery well designed, the lines rich, which shows of course that the cutting is good. Though there is no ornament save the beautiful initial S and the curious foliated initials, . . . the page is beautifully proportioned and stately. . . . The great initial S I claim to be one of the very best printers' ornaments ever made, one which would not disgrace a thirteenth century MS.' Cockerell's catalogue description of the book is included with this copy.

PLATES XXIII, XXIV

31

[ULM] SEELEN-WURZGARTEN
Conrad Dinckmut, 26 July 1483. f°. 17/133 woodcuts. [PML 199, Goff s-364]

The *Seelen-Wurzgarten* is an anonymous devotional tract filled with many exempla. Several earlier editions were printed, but Dinckmut's was the first with illustrations, which he reused, in many fewer repetitions, in later editions of October 1483 and December 1488. There were also at least three Augsburg editions, 1484 and after, all using a generally independent series of cuts.

PLATE XXV

32

[ULM] TERENTIUS. EUNUCHUS
[German] Conrad Dinckmut, 1486. f°. 28 woodcuts. [PML 202, Goff T-108]

The German *Eunuchus* was apparently printed by Dinckmut at the expenses of the translator, Hanns Nythart. The woodcuts were designed by the anonymous artist who also illustrated Thomas Lirer's *Chronik*, of which Dinckmut printed three editions. Morris wrote that the *Eunuchus* illustrations 'all have backgrounds showing (mostly) the streets of a mediaeval town, which clearly imply theatrical scenery; the figures of the actors are delicately drawn, and the character of the persons and their action is well given and carefully sustained throughout. The text of this book is printed in a large handsome black-letter, imported, as my friend Mr. Proctor informs me, from Italy. The book is altogether of singular beauty and character.'

PLATE XXVI

33

[BASEL] SPIEGEL DER MENSCHLICHEN BEHÄLTNIS
Bernhard Richel, 31 August 1476. f°. A set of early impressions of 254/257 woodcuts, without text. [PML 224, Goff s-664]

One of the most interesting items in Morris's library was this set of what have been called, for want of a better term, proof impressions of the illustrations for Richel's 1476 German edition of the *Speculum humanae salvationis*. With only minor exceptions these cuts, printed four to a page, appear in the same order as in the published edition. Two of the scenes are represented by different woodcuts from those used in the book, though all cut by the same hand. These woodcuts were pulled on the same stock of Basel-make paper as was used to print a large section of the full edition. Bound with the cuts is a manuscript of Johannes de Utino, *Compilatio librorum historialium totius Bibliae*. This was written on the same Basel-make paper, and its scribe may have written the German captions accompanying the cuts.

PLATES XXVII, XXVIII

34

[MAINZ] BREYDENBACH. PEREGRINATIO IN TERRAM SANCTAM
Erhard Reuwich, 11 February 1486. f°. 16 woodcuts + 7 folding views. [PML 30, Goff B-1189]

Erhard Reuwich, an artist originally from Utrecht, accompanied Bernhard von Breydenbach, dean of the Cathedral church of Mainz, on a pilgrimage to the Holy Land. On their return, Reuwich had some of his drawings made on the trip translated into woodcut, producing thereby the first illustrated travel book. The book uses Peter Schoeffer's types, but Reuwich stated explicitly that it was printed in his own house. Spaces left in the text show that more woodcuts were planned than actually used. Morris wrote that the *Peregrinatio* 'amongst other merits, such as actual representations of the cities on the road to the Holy Land, must be said to contain the best executed woodcuts of the Middle Ages.' The folding view of Venice is more than five feet long. After the Latin edition, Reuwich printed editions in German and Dutch. His woodcuts were subsequently used in editions printed in Lyons, Speyer, and Zaragoza, and other editions contain copied woodcuts. Morris's friend Fairfax Murray formed a very complete collection of early editions of the *Peregrinatio*.

PLATE XXIX

35

[NUREMBERG] STEPHAN FRIDOLIN. DER SCHATZBEHALTER
Anton Koberger, 8 November 1491. f°. 92/97 woodcuts. [PML 178, Goff s-306]

Stephan Fridolin was a Franciscan of Nuremberg; his *Schatzbehalter* or Treasure Chest was an enormous, complex devotional work organized around the Passion story. Its large and impressive woodcuts, for some of which drawings survive, have been attributed to Dürer's teacher, Michael Wohlgemut. They are very closely related to Fridolin's text, the mystical elements in particular being represented with a striking literalness. As happened not infrequently, the ambitious scheme of illustration was not completed, the main series of cuts stopping half-way through the book. Morris showed slide illustrations from the *Schatzbehalter* in his lecture on 'Woodcuts of Gothic Books,' remarking that 'Although so late, there is no trace of any classical influence in the design. The architecture, for instance, is pure late German architecture.'

PLATE XXX

36

[LYONS] BOCCACCIO. DE LA RUINE DES NOBLES HOMMES ET FEMMES

Mathias Huss & Johannes Schabeler, 1483. f°. 9 woodcuts. [PML 600, Goff B-712]

Mathias Huss was the kinsman, probably the brother, of Martin Huss, to whose Lyons press he succeeded in 1482. This French version of Boccaccio's *De casibus virorum illustrium* was apparently reprinted from Colard Mansion's splendid 1476 edition. However, the crude woodcuts are independent of the engravings found in a few copies only of that edition. The woodcut capitals, incorporating human faces and animal figures, seem to be the first use of this style of ornament in Lyons printing. Cockerell's catalogue note is included in this copy.

PLATE XXXI

37

[LYONS] BERTRAND DU GUESCLIN

Guillaume Le Roy, c. 1487. f°. 21/29 woodcuts. [PML 605, Goff G-541]

This rare edition of the military feats of Bertrand Du Guesclin (see also no. 40) was one of the late publications of Guillaume Le Roy, first printer in Lyons. Apart from the large portrait with arms of Du Guesclin, its woodcuts are a mélange of battle scenes taken from several earlier Lyons imprints; one series in particular is very primitive, with stiff, unshaded knifework.

38

[LYONS] LES QUATRE FILS AYMON

Jean de Vingle, 20 April 1493. f°. 21/28 woodcuts. [PML 611, Goff A-1433]

Les quatre fils Aymon was a popular prose tale of four rebel brothers, taken from a twelfth-century *chanson de geste*. There are at least seven fifteenth-century editions in French, all rare, six of them from Lyons. The present edition, known only from this copy, is the third, but the first with extensive illustration. The woodcuts, made specifically for this book, are very attractive. Jean de Vingle must have had good success with them, as he printed at least three more editions with essentially the same cuts.

PLATE XXXII

39

[ABBEVILLE] AUGUSTINUS. DE LA CITÉ DE DIEU

Pierre Gérard & Jean Du Pré, 24 November 1486 – 12 April 1486/7. 2 vol., f°. 23 woodcuts. [PML 622, Goff A-1247]

This large-folio edition of St. Augustine's *City of God* can lay strong claim to being the most splendid fifteenth-century French imprint. It is one of only three books printed at Abbeville in the fifteenth century, two of which were extraordinary productions. Jean Du Pré, whose name appears only in this one of the three, was a Parisian printer during these same years. It seems likely that he financed Gérard's shop—the type of the Abbeville books is his—but did not himself work there. The beautiful woodcuts were also almost certainly made in Paris and they returned there, showing up later in the stock of Antoine Vérard.

PLATE XXXIII

40

[ABBEVILLE] LE TRIUMPHE DES NEUF PREUX

Pierre Gérard, 30 May 1487. f°. 11 woodcuts. [PML 623, Goff T-458]

The anonymous *Triumphe des neuf preux* was dedicated to the reigning king, Charles VIII. The last chapter of the work concerns Bertrand Du Guesclin (d. 1380), the great Constable of France, whom the author equiparates with the nine traditional Worthies. Most of the Worthies are depicted in a striking, very vigorous wide-legged stance. The portrait of Du Guesclin is less conventionalized, and Morris once told Pollard 'that he was sure this bullet-head must have followed a true tradition of the living man, and a comparison with an authentic portrait has shown that this surmise was correct.' The provenance of the book has an impressive ring: Gaignat – Girardot de Préfond – Revoil – d'Essling–Seillière – Crawford of Lakelands – William Morris.

PLATE XXXIV

41

[PARIS] HONORÉ BONNOR. L'ARBRE DES BATAILLES
Antoine Vérard, 8 June 1493. f°. 25/116 woodcuts. [PML 506, Goff B-1023]

The third edition, but first with illustrations, of a very important late medieval treatise on warfare. Its publisher, Antoine Vérard, originally a calligrapher and illuminator, was the most enterprising figure in the fifteenth century Parisian book trade. He maintained an extensive stock of woodcuts which were made to serve a variety of uses. Many of the *Arbre des batailles* cuts had already been employed in other of his imprints. Curiously, just two weeks later another illustrated edition of the *Arbre des batailles* appeared in Paris. Its printer, Jean Du Pré, used woodcuts originally found in the great 1488–89 *Mer des histoires*.

42

[PARIS] PARIS ET VIENNE
Jean Tréperel, 1498. 4°. 11/27 woodcuts.
[PML 534, Goff P-113]

Paris et Vienne was a popular romance written in the 1430s and appearing in more than a dozen printed editions, in four languages, before the end of the century. Almost all of these survive only in single copies. The present unique copy is probably the fourth or fifth edition. The printer, Jean Tréperel, specialized in small, popular French-language texts, many of which are extremely rare. The cuts in his *Paris et Vienne* were with one exception made specifically for this edition, and are very attractive. In Pierpont Morgan's Catalogue the compiler, who for this section would have been Pollard or Proctor, says the volume 'was a great favorite of William Morris's.'

PLATE XXXV

43

[PARIS] BOCCACCIO. DE LA GÉNÉALOGIE DES DIEUX
Antoine Vérard, 9 February 1498/9. f°. 29/44 woodcuts. [PML 536, Goff B-755]

The only French, and only illustrated fifteenth-century edition of this text. It is a good example of Vérard's most elaborate, large-folio publications. At least four distinct series of woodcuts appear in the Boccaccio, most or all of which were originally made for other texts. The large mythological cuts are taken from Vérard's 1494 *Bible des poètes*, or paraphrased Metamorphoses, and are copies of Colard Mansion's 1484 Bruges edition of the same text.

PLATE XXXVI

44

[KUILENBURG] SPEGHEL ONSER BEHOUDENISSE

Johann Veldener, 27 September 1483. 4°. 129/130 woodcuts. [PML 657, Goff s-660]

Purchased by Morris in January 1891; his letter to J. H. Middleton several days after the purchase explains the volume's importance: 'I have just bought a very fine and interesting book: Speculum Humanae Salvationis (in Dutch), Culembourg, Veldener, 1483. That says little; but the point of it is that it has in it *all* the cuts from the block-book Speculum (116) and 12 more seemingly of the same date. These are not recut, but are printed from the original blocks sawn in two down the columns of the canopies: some of these cuts are to my mind far away the best woodcuts ever done, and generally the designs are admirable: at once decorative, and serious with the devotional fervour of the best side of the Middle Ages. The date of the cutting you know is probably about 1430.' Since Morris wrote, the dates of the four semi-blockbook Utrecht editions of the *Speculum* to which he refers have been more accurately placed between c. 1466 and c. 1479. Veldener printed in Utrecht before moving to Kuilenburg, making his ownership of the blocks easily explained. He printed another abbreviated edition of the Dutch *Speculum*, apparently on the same date; it survives in only one copy.

PLATE XXXVII

45

[SCHIEDAM] J. BRUGMAN. VITA LIDWINAE

Anonymous press, 1498. 4°. 23 woodcuts. [PML 661, Goff B-1220]

The *Vita Lidwinae* was the first book printed in Schiedam; the name of the printer is not known, but he used types imported from Gouda. Lidwina was a local cult figure whose relics were venerated in the parish church. At the age of fifteen she took a bad fall while ice skating, and the remainder of her life was spent in bed, suffering a succession of terrible afflictions. She consumed almost no food or drink, and toward the end her main nourishment came from frequent reception of the Sacrament. Three Dutch-language editions of a different text of Lidwina's life had already been printed. This Latin edition was related to an attempt by the townspeople of Schiedam to have her canonized. Its woodcuts are unusually vivid and specific, and are very skilfully cut.

PLATE XXXVIII

Calligraphy

WILLIAM MORRIS'S CALLIGRAPHIC MANUSCRIPTS HAVE IN THIS century been widely dispersed into various museums and libraries, public and private, on two continents. It is thanks to the generosity of many institutions and individuals that in the present exhibit it has been possible temporarily to gather together in one place the most comprehensive selection to date of Morris's calligraphic work, including all of his most important decorated manuscripts. The only stage of his calligraphic career not represented is that of the 1856–57 period. Of the three single leaves recorded from this period, two are presently unlocated. The third, the Browning *Paracelsus* stanzas, is in the Huntington Library.

46

EYRBYGGJA SAGA, 1869
[Fitzwilliam Museum]

The text is Morris's translation of one of the first Icelandic sagas he read and studied with Eiríkr Magnússon. On pages 97–101 Morris's ordinary hand changes to a handsome upright script, with closely spaced lettering, which looks forward to that in his first calligraphic version of *The Dwellers at Eyr* (no. 48).

PLATE XLI

47

BELLEROPHON, from The Earthly Paradise, 1869
56 leaves. [British Library Add. MS.45301]

Bellerophon was one of the last poems written for *The Earthly Paradise*. Many of Morris's literary manuscripts of this period contain incidental marginal decorations and sketches, representing breaks in the flow of his rapid composition. In Bellerophon these decorations on several pages take the form of calligraphic letter designs. Some of the designs relate closely to the script of *The Dwellers at Eyr* (no. 48), and also to that of *The Volsungs and Niblungs* (no. 49), and the *Book of Verse* (no. 50).

PLATE XLII

48

THE DWELLERS AT EYR, c. 1869
50 pages (37–46 lacking). 14 × 9½ in.
[Bodleian Library MS. Eng. misc. c. 265]

This *Dwellers at Eyr* was Morris's first substantial decorated calligraphic manuscript of the 1869–75 period. The script has connections in specific letter forms with that in the fair-copied sections of the *Eyrbyggja* manuscript (no. 46). Like it, the *Dwellers at Eyr* was written before Morris began using a flat-nibbed pen, so that there are no strong contrasts of thick and thin lines. But the overall shape of the script is different from the *Eyrbyggja* script. It has cursive features which seem to indicate the influence of Morris's copy of Arrighi's writing book, though its letters are broader and more widely spaced than in the Italian chancery scripts. The chapter headings were added later, and are in a script almost identical to that in *A Book of Verse* (no. 50). The floral decoration is made up of discrete small sprays of flowers.

PLATE XLIII

49

THE STORY OF THE
VOLSUNGS AND NIBLUNGS, c. 1870
146 pages. 11 × 8½ in.
[Bodleian Library MS. Eng. misc. d. 268]

The Volsungs and Niblungs was probably Morris's second full calligraphic manuscript. The script was written with a flat-nibbed quill pen, lending strongly contrasted weights of strokes; it is a more regular script than in the first *Dwellers at Eyr* (no. 48). Certain of the letters —h, k, m, and n—carry distinctive thin diagonal strokes. On the first page the ornate border is made up of the same discrete floral elements as in the *Dwellers at Eyr*. Only one of the eight musicians in the border (probably drawn by Morris) has been colored. The miniature of Sigurd sitting on the dragon Fafnir was painted by Charles Fairfax Murray.

PLATE XLIV

50

A BOOK OF VERSE, 26 August 1870
55 pages. 8½ × 6 in.
[Victoria & Albert Museum, L.131–1953]

A Book of Verse was Morris's first dated calligraphic manuscript, and the first of four which he made for Georgiana Burne-Jones. Its script is very similar to that of *The Volsungs and the Niblungs*. The decoration of the pages is in some places so lush that the script is almost buried under intrusive foliage. One drawing in the manuscript is by Edward Burne-Jones; the other miniatures, including the portrait of Morris on the title, are by Charles Fairfax Murray. The ornament on the first ten pages was done by George Wardle; it follows the same system of regularly patterned floral sprays used by Morris in several other manuscripts (nos. 48, 49). After page 10 Morris took over the decoration, which changes to denser, more organic, almost self-generating floral arrangements.

PLATE XLV; COLOR PLATE LIV-a

51

THE DWELLERS AT EYR, 19 April 1871
249 pages. 14 × 9½ in. [Birmingham City Museums and Art Gallery 92'20]

This was the second of the manuscripts Morris made for Georgiana Burne-Jones. Its script is related to that of *A Book of Verse* (no. 50) but is even more ornate, with diagonal pen-flicks or serif strokes added to several ascenders, as in b and l. The first page is titled with a solid block of massed capitals, a style found in later calligraphic manuscripts and also in the Kelmscott Press books. Its border is particularly rich in its interweave of many differen-

flowers, fruits, and leaves. Over one hundred pages of the first manuscript of *The Dwellers at Eyr* (no. 48) were removed by Morris to complete the present manuscript.
PLATE XLVI

52

THE STORY OF KORMAK, 1871
42 pages. 15½ × 9¾ in.
[PML: gift of John M. Crawford, Jr.]

Kormak is one of a small group of calligraphic manuscripts written by Morris in two columns on large-format leaves in a roman script. A tentativeness in some of the pen-strokes shows that the hand was not fully mastered. Decorations are sketched in sepia or pencil, but none has been colored. Spaces were left for additional ornamentation. Bound in with this volume are other calligraphic trials: a page of a *Heimskringla*, two pages of *Hafbur and Signi*, and four pages of *Frithiof* (see no. 53). The Morris-Magnússon translation of *Kormak* was published for the first time, from this manuscript, in 1970 by the William Morris Society.
PLATE XLVII

53

THE STORY OF FRITHIOF THE BOLD, c. 1871
22 pages. 15½ × 9¾ in.
[Doheny Memorial Library]

Frithiof was the only two-column manuscript of Morris's in which we can see his completed plan of decoration for this layout. The early decoration was shared between Morris and Charles Fairfax Murray. Additional ornamentation and gilding was added to the manuscript in the twentieth century by Louise Powell and Graily Hewitt.
PLATE XLVIII

54

THE RUBAIYAT OF OMAR KHAYYAM, 10 October 1872
Vellum, 23 pages. 6⅛ × 4⅝ in.
[British Library Add. MS.37832]

This was the third manuscript made by Morris for Georgiana Burne-Jones. It is written in a small, unassertive roman script; considerable space is left between the lines. The decoration is thicker and more colorful than in *A Book of Verse*. There are five very heavily ornamented pages, four of which make up two double-page spreads. The foliage and flowers on these pages entirely fill the margins, giving a carpet-pattern effect. The figures in the manuscript were painted by Charles Fairfax Murray.
COLOR PLATE LIV-d

55

THE RUBAIYAT OF OMAR KHAYYAM, c. 1872–73
20 pages. 10¾ × 8½ in.
[Earl of Oxford and Asquith]

This second manuscript of the *Rubaiyat* written by Morris was done for Edward Burne-Jones, who painted its six miniatures. The remaining decoration is Morris's. Its color tones are darker than that of the first *Rubaiyat* (no. 54), but the volume of foliage is much less. The script is the same as in the first *Rubaiyat*, but with taller ascenders.

56

THE STORY OF THE YNGLINGS, c. 1872

Vellum, 97 pages. 10 × 8¼ in.
[Kelmscott Manor (Society of Antiquaries)]

The script is a variant of the small roman script found for instance in the first *Rubaiyat* (no. 54), with taller ascenders and a strong vertical emphasis. The a and e are in uncial form. The manuscript is extensively decorated; Charles Fairfax Murray painted the head of Odin, and Philip Webb contributed two rabbits in the margin of page 4. Acanthus leaves make an early appearance here, mixed with willow leaves; they were used later in the Horace (no. 62), the Virgil (no. 63), and in various Kelmscott Press borders.

COLOR PLATE LIV-b

57

THE STORY OF HALFDAN THE BLACK. THE STORY OF KING HARALD, c. 1872
24 vellum leaves, 9½ × 6¾ in.
[PML: gift of John M. Crawford, Jr.]

The script of *Halfdan* is almost identical with that of the *Ynglings* (no. 56), with the same uncial-form a and e, and also with small pen-flicks rising from the head of t. On the first page, titling in sepia may have been preparatory for further ornamentation. In the remainder of the manuscript titles and initials are lightly sketched in pencil. *King Harald* is incomplete here, its remaining leaves being bound in a composite volume of Morris's calligraphic work at the Bodleian Library (including no. 58). The *Halfdan* manuscript was presented to Lady Anne Blunt by Jane Morris in 1897.

PLATE XLIX

58

HAROON AL RASHEED, c. 1873

7 pages. 8¼ × 6 in.
[Bodleian Library MS. Eng. misc. d. 265]

The text of *Haroon al Rasheed* was taken from Jonathan Scott's *Tales . . . translated from the Arabic and Persian*, 1800, a copy of which was in Morris's library. Its script has a number of unique features compared with Morris's other calligraphic hands. It is heavy, very slightly slanted, with curved ascenders and a number of joined letters in the cursive manner. Many of its capital forms are very elaborate.

PLATE L

59

TRIAL SCRIPTS, c. 1874
Single leaf. [Norman H. Strouse]

These trials in various scripts are written on the reverse of a discarded Virgil leaf (see no. 63). The script at top, written upside-down, is almost identical with that of *Haroon al Rasheed* (no. 58). The next down, 'let us try some hands,' has many letter forms related to that of the *Virgil*, though its overall character is quite different. The next script is very similar to that in a fair-copy manuscript by Morris of *Lancelot du Lac*. At the bottom are lines copied from the title page of Arrighi's *Il modo de temperare le penne*, 1523. The script is copied from Arrighi's, but with thicker letters, shorter ascenders, and wider spacing.

PLATE LI

60

RUBAIYAT OF OMAR KHAYYAM, 1873-74
Vellum, 23 pages. 6 × 4½ in.
[Bodleian Library MS. Don. f. 3]

Morris's third *Rubaiyat* was of the same size as the one he did for Georgiana. Ornament appears only on the first two pages. The white vine ornament is characteristic of Morris's decorative technique at this period. The border decoration is closely related to that in the Horace (no. 62). Two distinct scripts are employed in the manuscript. The first is basically that of Morris's first two *Rubaiyats* (nos. 54, 55). The second is a more rounded roman with shorter ascenders.

PLATE LII

61

THREE ICELANDIC SAGAS, c. 1873-74
240 pages. 10 × 7¾ in.
[Fitzwilliam Museum MS.270*]

This collection of sagas was the fourth of the calligraphic manuscripts created by Morris for Georgiana Burne-Jones, whose initials he put in gilt at the bottom of the first page. The script is an angular italic joined with a variety of rather eccentric capitals, some with gothic complications and some quite awkward looking. At the opening of each of the sagas is a large initial with white vine decoration growing into the margin, beyond the confines of the initial's square.

PLATE LIII

62

HORACE, ODES, 1874
Vellum, 183 pages. 6⅝ × 4⅝ in.
[Bodleian Library MS. Lat. class. e. 38]

Horace was one of Morris's smallest calligraphic manuscripts. May Morris described it as 'a delightful volume (I use the word carefully), instinct with joy, vivid and jewel-like.' Due to its small size, its italic script is particularly condensed. The decoration is not complete, but substantially finished. Morris planned a very elaborate page to open each of the four books of Odes. He completed only that for book II, but the other books contain sketches for decoration.

COLOR PLATE LIV-c

63

VIRGIL, AENEID, 1874-75
Vellum, 370 pages. 13⅝ × 8¾ in.
[Doheny Memorial Library]

Morris planned the Virgil as his calligraphic masterpiece. Burne-Jones, who was enlisted to design miniatures for the manuscript, wrote that 'it was to be wonderful and put an end to printing.' Morris himself copied the first of Burne-Jones's scenes (the drawings for which are in the Fitzwilliam Museum) into the manuscript, and others were done by Fairfax Murray. The very handsome floriated initials are Morris's work. The script Morris used was obviously carefully chosen to suit the dignity and monumentality of the project. The shoulder line of the letters is very strongly emphasized with horizontal serifs, notably in u, v, and y, yielding a subtly archaic appearance. The overlarge g is the only obtrusive letter. The Virgil later came into the possession of Fairfax Murray, who commissioned Graily Hewitt to complete the text (Morris had written only 177 pages) and, with Louise Powell, to add decoration.

PLATE LIV; COLOR PLATE LIV-e

Printing and Book Design

THE STORY OF WILLIAM MORRIS'S INVOLVEMENT WITH PRINTING and book design may be characterized as a struggle for control of the means of production. Surely no previous printer, or none since Gutenberg, so effectively concerned himself with every step of the printing process as Morris did at the Kelmscott Press. The early projects in bookmaking were never completed, and from what remains it seems doubtful that they could in any circumstances have been totally successful; but they have been treated at some length as showing the seriousness of Morris's interest for many years before his Press was begun. Two later books, *The House of the Wolfings* and *The Roots of the Mountains*, deserve special notice as being simple but impeccably designed commercial publications. By conscious decision, only a few pages from Kelmscott Press volumes have been illustrated, the major emphasis being rather on original designs by Morris and Burne-Jones. The beauty of the Kelmscott Press books depends in great measure on the interaction between sharp, well-inked types and blocks, and strong, rough-surfaced paper. Almost the entire effect of this interaction is lost in photography.

Before Kelmscott

64

CUPID AND PSYCHE: THE EARTHLY PARADISE

The Earthly Paradise, Morris's longest work, was written over a period of years in the 1860s. It is a cycle of twenty-four stories in verse, set as a series of tales exchanged, two each month, by a group of Norse exiles and their Greek hosts. Perhaps as early as 1864 and certainly in 1865 Morris and Burne-Jones hatched a plan of publishing this great cycle in a lavishly illustrated edition, their first collaboration in joining story and picture. The medium of illustration would be wood engraving; there were to be, according to Georgiana Burne-Jones, 'two or three hundred woodcuts.' By September 1865, shortly before Morris left Red House, many were 'already designed, and some even drawn on the block.' Sydney Cockerell recorded that Burne-Jones finished designs for four of the tales, 'Cupid and Psyche,' 'Pygmalion,' 'The Ring Given to Venus,' and 'The Hill of Venus.'

Various drawings and sketches by Burne-Jones for all these survive.

The most advanced work was done for Cupid and Psyche, the first story in May and probably Morris's earliest *Earthly Paradise* composition. Substantial groups of Burne-Jones's drawings for 'Cupid and Psyche,' including many on tracing paper for transfer to blocks, are preserved in the Birmingham City Museums and Art Gallery, the Ashmolean Museum, and the Pierpont Morgan Library. Forty-four wood blocks of these scenes were bequeathed by May Morris to the Society of Antiquaries; from scattered surviving pulls, we know that several more at least were cut. Most of these engravings were done by Morris himself, and a few by others in his circle, after unsuccessful attempts to have the work done by commercial wood engravers.

Work on illustrations for *The Earthly Paradise* continued at least into 1867. But the ambitious plan for the book fell through, in chief part probably because of the impossibility of integrating the woodcuts appropriately with contemporary typography. The 'Cupid and Psyche' woodcuts were first published together with Morris's text in a limited edition in 1974 by Clover Hill editions, with an extensive introductory study by A. R. Dufty.

A BURNE-JONES'S DRAWINGS FOR THE EARTHLY PARADISE [PML]
This album of thirty-nine drawings formerly belonged to F. S. Ellis. All but three of the drawings are on tracing paper; twenty-one scenes from 'Cupid and Psyche' are represented, some with variant drawings. Not all these scenes were cut in wood. There is one drawing for 'Pygmalion' (see below), several for 'The Hill of Venus,' and several are un-identified. The drawing illustrated represents 'The maiden torch-bearers' unhappy bands,' elements of which appear in one of the completed woodcuts. $4\frac{1}{8} \times 6\frac{1}{4}$ in.,
PLATE LV

B BURNE-JONES'S DRAWING AND ENGRAVING FOR PYGMALION [PML]
The same album contains Burne-Jones's drawing of Pygmalion sculpting Galatea. With it, on Japan paper, is a pull of a copperplate engraving made from the drawing. The engraving is at best characterless, and Ellis has annotated: 'This drawing was made for an experiment of engraving on copper—the wretched engraving of which this proof only was taken was such a failure that no more drawings were made.' At least passing consideration must have been given to an idea of illustrating *The Earthly Paradise* with engravings, for two Burne-Jones engravings for 'The Ring Given to Venus' are also known. $4\frac{1}{8} \times 3$ in.
PLATE LV

C PSYCHE BEFORE PAN: WOOD ENGRAVING BY JOSEPH SWAIN
[PML: gift of John M. Crawford, Jr.]
From the account of George Wardle, who transferred Burne-Jones's drawings to the woodblocks, we know that some of the 'Cupid and Psyche' illustrations 'were given at first to "the trade" to be cut but the result was so unsatisfactory that Morris tried to get the cutting done by un-professional hands.' The present proof is apparently the only one known that can be correlated with this 'professional' experiment, for it is signed 'Swain sc.' Joseph Swain was a leading wood engraver of the 1860s, who rendered Burne-Jones's drawing in precisely the accurate but conventional

manner that Morris was trying to supersede. $6\frac{3}{8} \times 4\frac{5}{8}$ in.

PLATE LVIII

D SPECIMEN LEAVES FROM THE CHISWICK PRESS, C. 1866
[PML: gift of John M. Crawford, Jr.]
It was long known that sample pages of 'Cupid and Psyche' were set up by The Chiswick Press, and one of these pages was described in an anonymous article in the *Studio*, October 1898, but no recent Morris scholar had seen them. These are those long-lost pages, showing the illustrated *Earthly Paradise* project in print. Eight double-column folio pages (on two sheets of imitation laid paper) were set in Caslon using a long s; these include two-page spreads, showing how the large 'Cupid and Psyche' cuts were meant to head the columns in a frieze effect. Seven of the 'Cupid and Psyche' cuts appear on these pages. Renaissance-style woodcut initials were used from the extensive Chiswick Press stock cut by Mary Byfield. Two pages were also set in Basle Roman. Basle Roman was—and is—available in only one size; the titling was set in Caslon. Both samples have side-notes in italic. $14\frac{1}{8} \times 9\frac{1}{2}$ in. (Caslon leaves); $14\frac{3}{8} \times 9\frac{5}{8}$ in. (Basle leaf).

PLATES LVI, LVII

E THE EARTHLY PARADISE. 3 VOL. IN 6. F. S. ELLIS, 1868–70 [PML]
The published edition of *The Earthly Paradise* has a simple, plain but attractive appearance. The only decoration in the volumes was a pretty woodcut vignette of three musicians, designed by Burne-Jones and cut by Morris. The ordinary paper issue was printed in three volumes, bound in green cloth. The large- or fine-paper issue (only $2\frac{1}{2}$ cm. larger than the ordinary issue), in twenty-five copies, was printed on high-quality handmade paper, of C. Ansell make in the first four volumes, and of J. Whatman make in the final two. The thickness of this paper required that the work be bound in six volumes, which was done in neat holland boards. This copy was presented by William Morris to his mother.

PLATE LIX

F PULLS OF THE CUPID AND PSYCHE WOODCUTS
[PML: gift of John M. Crawford, Jr.]
At least two distinct early sets of pulls from the 'Cupid and Psyche' woodcuts are known. One set, probably in a very few copies, was made by friction printing in 1881, apparently by Morris himself. A second set was pulled by Emery Walker for Morris during the Kelmscott Press years. According to Cockerell, about eight copies were made, of which this is presumably one. These copies are all on Kelmscott paper.

PLATE LX

G CUPID AND PSYCHE TRIAL LEAF, 1897 [PML: gift of John M. Crawford, Jr.]
After Morris's death, Burne-Jones continued to think about the 'Cupid and Psyche' project: he proposed that an edition of Morris's text be printed using some of the old woodcuts 'with new engravings of the less successful subjects.' His own style had changed considerably in the last three decades, and it would not have been an easy matter to marry the old Burne-Jones idiom to the new. In January 1897 various trial leaves of 'Cupid and Psyche' were printed at the Kelmscott Press in Troy type, most of them heading the text with the

sixth 'Cupid and Psyche' woodcut, 'The Procession to the Hill.' The next month an agreement was made that Longmans would publish a limited edition of the book, to be printed at the Chiswick Press. Some Burne-Jones redrawings of 'Cupid and Psyche' scenes are known from this period, but the project did not advance rapidly, and it died with the artist in June 1898.

PLATE LXI

65

LOVE IS ENOUGH

In the autumn of 1871, after returning from his first trip to Iceland, Morris began a poetic romance, *Love Is Enough*. These were the years of his strong calligraphic interest, and from its first conception Morris envisioned his poem as appearing in an illustrated and decorated edition. Rossetti wrote in October 1871 that the book would have 'woodcuts by Ned Jones and borders by [Morris] himself some of which he has done really very beautifully.' Burne-Jones wrote several weeks later that 'it will come out some time next summer, and I shall make little ornaments to it.' The enterprise advanced in 1872 to the point of having trial pages printed with woodcut decoration, but like the *Earthly Paradise* project was finally aborted.

A TRIAL PAGES WITH ORIGINAL
   DECORATION [Dr. Gerald Wachs]
These two leaves, formerly belonging to George Campfield, foreman of Morris and Co., seem to be the only *Love Is Enough* trials known with original decoration by Morris. One page contains a roughed-in initial L, similar in design to the two that were cut.

There is also a rough sketch for a side border and a small horizontal border. On the following page is a finished design for a side border, which was cut.

PLATES LXII, LXIII

B TRIAL LEAVES WITH WOODCUT
   DECORATION
   [PML: gift of John M. Crawford, Jr.]
Two sets of the same two printed leaves discussed above. One has pasted-on woodcut decoration: a handsome floriated L, an apple border on the left, and a border of four putti, Burne-Jones's design, on the right. The other trial does not have attached decoration, but is accompanied by pulls of three borders. In all, eight border decorations were cut in wood, seven with floral patterns designed by Morris, plus the putti border of Burne-Jones.

PLATES LXIV, LXV

C LOVE IS ENOUGH. ELLIS & WHITE,
   1873 [Dr. Gerald Wachs]
As with the 'Cupid and Psyche' scheme, the weakness of the types against the woodcuts may have discouraged Morris. The published edition of *Love Is Enough* is unillustrated and conventional in appearance. Its one distinction is a beautiful trade binding designed by Morris: green cloth with a simple willow- and myrtle-pattern gilt frieze, related to Morris's manuscript decoration of the period.

D LOVE IS ENOUGH, LARGE-PAPER
   ISSUE. ELLIS & WHITE, 1873
   [Gordon N. Ray]
Twenty-five large-paper copies were issued on handmade Whatman paper. This might equally well be called a tall-paper issue, for the leaf width is slightly narrower than that of the

trade issue. Morris's three musicians vignette is used as a tailpiece in the limited issue (see no. 64e).

E LOVE IS ENOUGH, VELLUM ISSUE.
ELLIS & WHITE, 1873 [PML]

Four copies of *Love Is Enough* were printed on vellum. This copy has a presentation inscription from Morris to 'Georgie,' Georgiana Burne-Jones; it later belonged to A. N. L. Munby.

F LOVE IS ENOUGH, KELMSCOTT PRESS,
24 MARCH 1898 [PML]

The Kelmscott Press edition of *Love Is Enough* was the next-to-last publication of the Press. The tailpiece opposite the colophon, showing the coronation of the lovers, was engraved by Hooper after a design Burne-Jones made in 1872 as the frontispiece for the then-planned illustrated edition. Shown with the volume is a platinotype of Burne-Jones's drawing, retouched with ink and china white (gift of John Crawford).

66

A DREAM OF JOHN BALL AND A KING'S LESSON
Reeves and Turner, 1888
[PML: gift of John M. Crawford, Jr.]

A Dream of John Ball was published in April 1888. It has an etched illustration by Burne-Jones, one of his few pieces of book work before the Kelmscott Press was founded. In the Kelmscott Press edition of John Ball, 1892, the illustration was reworked in woodcut, with lettering designed by Morris. This large-paper copy of the first edition, on John Dickinson handmade paper, has a presentation inscription from Morris to Burne-Jones, March 31, 1888. It later belonged to Hugh Walpole. John Crawford's gift includes a second large-paper copy presented by Morris to his daughter Jenny, and later given by May Morris to Sydney Cockerell.

PLATE LXVI

67

A TALE OF THE HOUSE OF THE WOLFINGS
Reeves and Turner, 1889
[PML: gift of John M. Crawford, Jr.]

The House of the Wolfings, printed in the autumn of 1888 at the Chiswick Press, bears the first fruits of Morris's growing fascination with typography. The text is in Basle Roman, which was chosen after consultation with Emery Walker. The title page is extremely effective. The titling proper is in Caslon capitals and the accompanying verse in Basle capitals. Buxton Forman tells the story of being at the Chiswick Press when the author was overseeing the title. Morris insisted on adding a superfluous word to one of the lines to improve its visual balance. When Buxton Forman mildly protested, he responded, 'What would you say if I told you that the verses on the title-page were written just to fill up the great white lower half? Well that was what happened!' One hundred copies were printed on John Dickinson handmade paper, with an inserted limitation slip. The bold title page sits well on the large paper, but the text is incongruously small against the enormous margins. Initially Morris took great pride in the appearance of this book, speaking of it as 'a pretty piece of typography for modern times.' A short while later, though still generally pleased, he saw flaws; he agreed with F. S. Ellis that the printers had 'managed to knock the guts out of [the type] somehow. Also I am beginning to learn something about the art of type-setting; and I now see what a lot of difference there is between the work of the conceited numskulls of to-day and that of the 15th and 16th century printers merely in the arrangement of the words, I mean the spacing out: it makes all the difference in the beauty of a page of print.'

The present large-paper copy has a presentation inscription from Morris to Burne-Jones, 6 January 1889. Also in the Morgan Library is a large-paper presentation copy to Philip Webb.

PLATE LXVII

68

THE ROOTS OF THE
MOUNTAINS
Reeves and Turner, 1890
[PML: gift of John M. Crawford, Jr.]

The Roots of the Mountains was printed at the Chiswick Press and has the same general design as *The House of the Wolfings*, with a number of small changes. The lowercase e in the Basle fount was replaced. The original Basle e has a slanted bar, in the style of fifteenth-century romans; Morris had a more conventional e with horizontal bar substituted. He lengthened the title so that it would run over to seven lines, as compared to the five-line title of *The House of the Wolfings*. The numbers on the title page are roman rather than arabic. The chapter titles are set run-on rather than balanced, and here too Morris seems to have increased their length so as to spill them onto two lines or more. There are side-notes instead of a head-line. Despite his earlier protest about the compositors of *The House of the Wolfings*, the setting in *The Roots of the Moun-*

tains is not noticeably more compact. Morris had by now seen the error of his ways concerning overly large paper. A special make of Whatman handmade paper was bought to print 250 copies, and the type-page was designed to fit this size. However, the monumental title page could comfortably use more margin than it has been given even on the larger paper. The Whatman paper copies were bound in Morris & Co. chintz. Morris declared *The Roots of the Mountains* to be 'the best-looking book issued since the seventeenth century.'

PLATE LXVIII

69

THE SAGA LIBRARY
6 vol. Bernard Quaritch, 1891–1905
[John A. Saks]

The Saga Library reprints the bulk of the joint Icelandic translations by Morris and Eiríkr Magnússon; the Volsunga Saga was originally to have been part of the series, but was later omitted. Morris's agreement of publication for the volumes, dated 3 July 1890, stipulated that he was 'to have a control of the choice of type, and the style of printing of his Sagas.' This may be compared with his agreement with Quaritch two months later for *The Golden Legend* (see no. 79b). Quaritch seems to have expressed some concern as to just what this control involved, for ten days later Morris wrote him to say that 'the size & kind of type which I think would suit the Saga Library best would be Caslon's Pica: the next size smaller would bring you to the same as the Earthly Paradise.' He also gave his opinion 'that the large paper copies should be printed with the hand-press, so as to get the impression black enough.' The design of the Saga Library is simple and unpretentious, but there are some indications, as in the run-on chapter headings, of Morris's influence. Caslon was not used, being replaced by its nineteenth-century rationalized variant, Old Style. Some minor improvements of design were made in volume 3, the first volume of the Heimskringla (1893). The binding, in quarter green morocco and green cloth, with gilt willow ornaments, was designed by Morris.

This copy of *The Saga Library* contains presentation inscriptions from Morris to Sydney Cockerell, 15 February 1896, in the first three volumes. Included are several proof pages from the Chiswick Press, with corrections by Morris, and an A.L.S. from Morris to Eiríkr Magnússon.

70

THE STORY OF GUNNLAUG THE WORM-TONGUE AND RAVEN THE SKALD
Chiswick Press for William Morris, 1891

Morris's translation with Eiríkr Magnússon of the Gunnlaug Saga, first published in 1869, was printed at the Chiswick Press in seventy-five copies, on leftover paper from the Whatman stock used for *The Roots of the Mountains*. Three additional copies were printed on vellum. This small volume was never formally issued, even privately, and must have been pure experiment. A number of copies were given away as keepsakes by Mrs. Morris after her husband's death. The type is a facsimile of Caxton's fourth type, supplied to the Chiswick Press, like Basle Roman, by William Howard. It appeared in a Chiswick Press specimen in 1867, and *Gunnlaug* was its only use in a full book. Though dated 1891 in the colophon, Sydney Cockerell records that the printing was finished in November 1890, at the time Morris's Golden type was nearing completion.

A TRIAL LEAVES [British Library]
The trial leaves for *Gunnlaug* preserve several early variants of composition, one with so narrow a set that a two-column printing might almost have been contemplated. Initials were left blank for hand decoration, and it is said that Morris decorated several copies. On one of the trial leaves is an unfinished initial, presumably by Morris, in pencil, red ink, and blue watercolor.
PLATE LXIX

B THE FINISHED BOOK
[PML: gift of Miss Elizabeth Reilly]
The present copy must have received its decoration later, by someone outside the Morris circle, though initials are copied from Kelmscott Press initials. There is also an undecorated copy in the Morgan Library. Despite its archaic design, *Gunnlaug* betrays its Victorian origin in one particular: the 'numskull' compositor (see no. 67) spaced out the lines much more openly than Morris would have approved. This fault was remedied in the Kelmscott Press books.

Founding the Kelmscott Press: Types and Paper

71

THE GOLDEN TYPE

Morris began to design his Golden type late in 1889. He referred to this type over the next months as Jenson-Morris, thereby accurately summing up its genealogy: it was derived from the fifteenth-century roman type of Nicholas Jenson, as remodeled by William Morris. The remodeling was extensive. Sydney Cockerell tells us that two Venetian incunabula were acquired by Morris about this time, specifically to study their types: Jenson's 1476 edition of Pliny in Italian, and Jacobus Rubeus's 1476 edition of L. Aretino's *Historia Fiorentina*. Both these volumes appear in the 1890–91 catalogue of Morris's library. To forestall confusion, it should be said at once that the type in both these books is Jenson's. Rubeus, or Le Rouge, was a friend and fellow-countryman of Jenson, and may have started out in his shop. Six of the seven types Rubeus used in Venice are very closely related to Jenson's types. His large roman, with which we are concerned, seems in fact to have been made from Jenson's punches or matrices, with the addition of two variant sorts; there is a slight difference in body size. The changes Morris wrought on this classic fifteenth-century type are discussed in greater detail below. His natural sympathy lay closer to gothic than to roman design, and this shows up clearly in his modifications. Some of Morris's ambivalence showed through in his note on his 'Aims in Founding the Kelmscott Press,' where he wrote that 'by instinct rather than by conscious thinking it over, I began by getting myself a fount of Roman type.' He told Walker that the Golden type had given him greater trouble than anything else he had attempted to design. The punches of the Golden type, as of the later Troy and Chaucer, were cut with great skill by Edward P. Prince, a freelance punchcutter recommended by Emery Walker.

A PHOTOGRAPHIC ENLARGEMENTS OF INCUNABULA
[St. Bride Printing Library]

Emery Walker made 5x enlargements of a number of incunabula to aid Morris's study of their types. This set of enlargements was given by Morris to Talbot Baines Reed, of Sir Charles Reed & Sons, in 1891. One of the enlargements is from a page of the Rubeus *Historia Fiorentina*.

B PHOTOGRAPHIC REDUCTION OF A PRELIMINARY DESIGN
[PML: gift of John M. Crawford, Jr.]

This photographic reduction was made from a drawing or retouched tracing by Morris of the same Rubeus page mentioned above. Many of the Jenson letterforms have been modified

in the direction of the final Golden type design: the movement toward regularized slab serifs, replacing Jenson's subtler brackets and swelled lines, has clearly begun. There is an interesting experiment, eventually abandoned, with a graceful long s. 3⅜×2⅝ in.
PLATE LXX

C PROCESS BLOCK REDUCTION OF A
   PRELIMINARY DESIGN
   [PML: gift of John M. Crawford, Jr.]
This process block was reduced from an enlargement of a different page of the Rubeus *Historia Fiorentina*. It is difficult to tell whether the thickening of the letters results from Morris's retouching of the type, or from the photographic process. The block may have been made to produce the weight of letter Morris was striving for. 3¼×3 in.
PLATE LXX

D ENLARGED PRELIMINARY DRAWING
   [PML: gift of John M. Crawford, Jr.]
This is pure drawing rather than tracing, but again the text is derived from the Rubeus *Historia Fiorentina*, and the scale is that of the photographic enlargements. The letters, all lowercase, are drawn in red ink outline, then filled in with india ink. Many modifications of the Jenson face have been made, most generally in the addition of thicker serifs. As in 71b, the minute dots to the i in the Jenson face have been replaced by the diamond shape that Morris favored. But he still positions the dots to the right, whereas in the final version they are centered. 10½×12¼ in.
PLATE LXXI

E PHOTOGRAPHIC REDUCTION OF A
   LATER DESIGN
   [PML: gift of John M. Crawford, Jr.]

Several variant forms, notably of the g and s, can be seen. There are also designs for w, which does not appear in the Jenson fount. 3½×3 in.
PLATE LXXII

F SMOKE PROOFS AND PHOTOGRAPHS OF
   SMOKE PROOFS
   [PML: gift of John M. Crawford, Jr.]
May Morris has described how, as Prince's exacting task of cutting progressed, her father would carry about the latest smoke proofs in a match box to examine in spare moments. The photographic enlargements represent two punches of the letter h, the first letter cut. The first, rejected by Morris, is close to the Jenson h, preserving even the slight rightward swelling at the top of the stem. The second, further modified to Morris's instruction, caps the stem with a slab serif and has a flatter curve in its shoulder. Smokes of b and the second h were impressed together on one slip, and of eleven uppercase letters on another slip.
PLATE LXXII

G EARLY TRIAL OF GOLDEN TYPE
   [PML: gift of John M. Crawford, Jr.]
At the time these sample lines were set only eleven lowercase letters had been cut. This was presumably the specimen of which Morris sent an 'overinked' example to F. S. Ellis on 27 August 1890. An earlier and even more nonsensical specimen is known, pulled when only five lowercase letters, a, b, e, h, t, had been finished. 2⅛×3⅞ in.
PLATE LXXII

H THE PROGRESS OF THE CUTTING
   [PML: gift of John M. Crawford, Jr.]
A note in Morris's hand of the lowercase let-

ters still to be cut, with a rough sketch for uppercase R. By mid-October all the lowercase letters were cut, and Morris began sample printing in January 1891, when the uppercase E and N were not yet ready (see no. 77). 7×4⅜ in.

PLATE LXXIII

72

THE TROY TYPE

On the same day he showed Emery Walker his first sample page in Golden type, Morris, in his exultation, spoke of designing a new black-letter fount. He later said that his goal was 'to redeem the Gothic character from the charge of unreadableness which is commonly brought against it.' The project was delayed slightly, but in the summer of 1891 Morris began work on the new type.

A EARLY DESIGN [Sanford L. Berger]
Unlike the Golden type, the new gothic type had no specific model. The present sheet shows two early stages of design. Four states of the lowercase have been drawn, each with many variants from the others. Virtually all the uppercase letters were changed before the punches were cut, but in weight and proportion these early sketches are nonetheless very similar to the finished fount. 9⅞×15½ in.

PLATE LXXIV

B LOWERCASE, FINAL DESIGN
   [Sanford L. Berger]
This sheet contains Morris's design of the lowercase of Troy type in very nearly final form. There are many small differences from the earlier sketch above, most strikingly apparent in f, g, k, and s. The final designs for the new alphabet were handed over to Edward Prince in August and September 1891, the lowercase being completed first. Some small variants between the design and the finished type result from the translation of oversize, inked letters to small metal punches, but there are no changes of conception. 10½ ×16½ in.

PLATE LXXV

C TROY TYPE SPECIMEN
   [Sanford L. Berger]
This specimen of twenty-six lines from Chaucer's 'Franklin's Tale' was composed in November 1891, when the Troy type was nearing completion. The fount in this state had an uppercase I with regular, symmetrical serifs, which Morris was displeased with, and discarded. The final-form I is less regular, with short spurs on the left of the upper serif and on the right of the lower. Morris's early conception of the type (no. 72a) had likewise envisioned an irregular I. 11¼×7⅞ in.

PLATE LXXVI

Though gothic letter persists in looking strange to modern eyes, Morris's Troy type was a great achievement, and his most important contribution to letter design. It was a natural outgrowth of his instinctive taste; 'to say the truth,' he wrote, 'I prefer it to the Roman.'

73

THE SUBIACO TYPE
Trial setting
[PML: gift of John M. Crawford, Jr.]

The Chaucer type was a smaller, pica (12-point) version of Troy, which Edward Prince began cutting in February 1892, only a few months after finishing off the larger version. In June 1892 Morris wrote Prince that in three months he hoped to 'be ready with a new set of sketches for a fount of type on English body,' the same size, that is, as Golden. What Morris might have had in mind for this type we do not know. On 5 November 1892, however, in the first days of cataloguing the Kelmscott House library, Cockerell recorded that Morris had out a copy of the Subiaco *De civitate dei* (1467), and had begun designing a lowercase after its model. Cockerell's later recollection was that Morris abandoned the project without having punches cut, but the present specimen shows that an entire lowercase was cut, in 16-point size. Morris's design was a version rather than a copy of the Subiaco type, with notably shorter descenders and many other variations. Certain letters—y, z, and g—are entirely unrelated to the Subiaco fount, and seem to go back to the early Troy type designs. The g, in particular, is identical to that on the bottom line of no. 72a above. Morris greatly admired the semi-roman Subiaco type, and expressed wonder that Sweynheym and Pannartz so soon abandoned it for a pure roman. Others of his circle also admired it. Several years later St. John Hornby decided to have a private type cast for his Ashendene Press. He consulted with Sydney Cockerell, Emery Walker, and Robert Proctor, and their decision was to model the type—cut by Edward Prince—on the Subiaco face. 4¼ × 3¼ in.

PLATE LXXVII

74

PUNCHES AND MATRICES OF THE KELMSCOTT PRESS TYPES
[Cambridge University Press]

The punches of all the Kelmscott Press types were cut by Edward P. Prince. The matrices were struck and types cast by the firm of Sir Charles Reed & Sons. Following the Kelmscott Press commissions, Prince continued to do ordinary commercial work, but he also became the chosen punchcutter for a long series of proprietary types, including those used by the Doves, Vale, Ashendene, Essex House, Eragny, Merrymount, Cranach, and Zilverdistel presses.

PLATE LXXVII

75

TYPES USED AT THE KELMSCOTT PRESS
9⅞ × 9¼ in.
[PML: gift of John M. Crawford, Jr.]

This specimen was printed on a cheap piece of proof paper, and must have been casually done. It gives an interesting conspectus of the three types, and shows exactly how many punches E. P. Prince cut for each: 83 for Golden, 91 for Troy, and 84 for Chaucer.

PLATE LXXVIII

76

JOSEPH BATCHELOR'S PAPER

Morris's concern with paper ran apace with his interest in typography in the months before the Kelmscott Press came into operation. Already, for *The Roots of the Mountains*, published in 1890, Morris had ordered the printing of a fine-paper issue on a good quality Whatman paper (no. 68). But he required an even better paper for the Kelmscott Press. Emery Walker recommended Joseph Batchelor, a hand papermaker of Little Chart, Kent, whose mill he and Morris visited in October

1890. As a sample of what was wanted Morris brought, in his words, 'a Bolognese paper of about 1473.' Morris's only Bolognese book of so early a date was the 1472 *Specchio di coscienza* of Antoninus Florentinus, and this may have been the model. The paper of the *Specchio* (as of many other Bolognese incunabula) does in fact bear a family resemblance to the Kelmscott papers: relatively thin and flexible but tough, with narrowly spaced wire-lines and not overly prominent chain-lines. Indeed, the character Morris wanted in his paper was general, as he realized, to a good many high-quality Italian fifteenth-century papers. The day after his visit to Little Chart he wrote Batchelor, 'I have looked into the various Venetian books of the fifteenth century and find that all the fine ones are printed on the same kind of paper as that I left you.'

From Morris's specifications Batchelor produced a superlative make of pure linen paper, well sized and very tough, ideally suited for slow, careful hand-press work. Three distinct stocks of paper were eventually made for the Kelmscott Press, all with watermarks after Morris's design and bearing his initials. The first was marked with a primrose, two marks being positioned in opposite corners of the sheet. The primrose paper was a 16×11 inch sheet, but made in two-sheet moulds. This was soon replaced with a double-size sheet of 16×22 inch, to allow *The Golden Legend* to be imposed more efficiently as a quarto. Its mark is also a primrose, with marks placed in all four corners of the sheet. Batchelor's charge for his paper was 2s per pound (plus the cost of making the moulds), which was some five or six times the price of ordinary machine-made book paper. A slightly larger sheet of about $16\frac{3}{4} \times 23$ inches was made for the Chaucer and was used also in many of the large quartos after 1892. Its watermark is a perch with a sprig in its mouth, a single mark being placed, conventionally, in the center of one-half of the sheet. The last stock of paper, made in 1895, was approximately 18×13 inches in size. Its mark is a simple apple on a two-leaved stem, with a mark centered in each half of the sheet. In 1895 Morris gave Batchelor permission to call this paper 'Kelmscott Hand Made,' to distinguish it from imitations.

PLATE LXXIX

77

FIRST PAGE PRINTED AT THE KELMSCOTT PRESS
[PML: gift of John M. Crawford, Jr.]

On 3 January 1891 (in a letter given by John Crawford to the Morgan Library) Morris invited a retired printer, W. H. Bowden, 'to engage yourself to me as compositor & press-printer in the little typographical adventure I am planning.' A cottage was rented at No. 16, Upper Mall, to house the equipment. Sir Charles Reed & Sons, typefounders, supplied a quantity of Golden type, and also galleys, composing sticks, quoins, and other materials. Bowden himself may have brought at this time a Demy Albion press which continued to be used in the shop for proofing. Bowden recalled that when the type arrived from Reed's, Morris pitched in to the work of laying it out in the cases; and when, as happened more often than not, he made a mistake, he would say to himself, 'There, bother it; in the wrong box again!' The next week two trial pages were printed, the first apparently being done on 31 January. A few initials had already been designed by Morris and cut by George Campfield, the foreman of Morris & Co. The uppercase E and N were not yet cast, so that in one of the trial pages, THEN is printed as THFT. Morris found the lowercase g to be too thick, and it was discarded in favor of a lighter version. The text of the trial pages was his recently finished romance *The Story of the Glittering Plain*, and this became, several months later, the first production of the Kelmscott Press.

The trial pages were printed on paper supplied just a few days before by Joseph Batchelor. Batchelor sent two small lots, one marked S, and the other H (the initials may refer to the softness or hardness of the sizing). The paper, he wrote Morris, was 'quite usable and is Antique, but is not so like the Venetian you left with me as I wish. . . .' The paper of the trial leaves is indeed slightly but noticeably different from the finished Kelmscott paper, being rather thinner and more flexible, with a softer surface. The day after the first page was printed, Morris, in high spirits, brought it over to show Emery Walker. They compared it 'with a page of the Saga Library which looked very poor beside it.'
PLATE LXXX

The Press Under Way

78

ORNAMENT: BORDERS, FRAMES, AND INITIALS
[PML: gift of John M. Crawford, Jr.]

Almost every single piece of decoration in the Kelmscott Press books—there are a few minor exceptions—was designed by Morris. Sydney Cockerell made a count in 1897 and, adding together the initials, borders, frames for woodcuts, other side ornaments, woodcut title pages, large initial words, and printer's marks, came to a total of 644 designs. This profuse variety is not immediately apparent on looking at the books because of the minor differences in many of the decorations. The initial T, for instance, always appears in a graceful Lombard form, but Morris designed thirty-four variants of the letter with differing floral backgrounds. John Crawford's gift to the Morgan Library includes many examples of Morris's original designs for Kelmscott ornament, only a few of which are shown here; other examples were already in the Library.

A DESIGNS FOR FOUR INITIALS
Here, Morris designed three initial I's side by side; their numbering is in Cockerell's hand.
PLATE LXXXI

B WOODCUT VERSUS ELECTROTYPE
The same four initials were cut together on one block of wood. Emery Walker has annotated this trial, 'one row printed from wood, the other from electrotypes done to convince W.M. that electro's cd. be used without artistic loss.' Walker gave a fuller account of this in an addendum to Catterson-Smith's note on photographing the Burne-Jones drawings for Chaucer (no. 96): 'when I suggested to William Morris the use of electrotype duplicates of the woodcut initials for use in the Kelmscott Press books, he with his suspicion of modern inventions was doubtful of their adequacy. I therefore made electrotypes from three woodcut letters and printed them upon the same piece of paper as the woodcuts and then asked Morris to say which was which. He put his finger on the electrotypes, but ad-

mitted that it was almost by chance, and said if there was no more difference than that we might use electrotypes. So they were used both for initials and other ornaments. The wood-engravings of pictures in the Chaucer and other books with Burne-Jones illustrations were printed directly from the wood blocks.'
PLATE LXXXI

C PROOFS OF INITIALS AND ORNAMENTS
According to a note by Cockerell this is one of three sets of pulls from the original woodblocks of Kelmscott initials and ornaments, made just before the Press closed. Because electrotypes were used in the Kelmscott Press books, there was no need to cut the initials free from the blocks they were engraved on. These proofs therefore give valuable information as to which decorations and letters were designed and cut together.
PLATE LXXXII

D PROOF OF INITIALS
This proof is annotated by Morris, 'the white borders to the letters want all opening out especially those with a cross x.' Three separate blocks of initials were proofed here. The roughed-out H and Q in the fourth row were never finished.
PLATE LXXXIII

E EXPERIMENTAL COLORED INITIALS
Morris had these blocks printed in blue, green, and black, then colored the letters themselves in red to test the effect. The Kelmscott edition of Wilfrid Scawen Blunt's *Love-lyrics* had initials printed in red at Blunt's request, but Morris was of two minds about their appearance. He wrote to his daughter Jenny that Blunt's volume was 'very gay and pretty with its red letters, but I think I prefer my own style of printing.' The *Laudes beatae Mariae virginis* (no. 87) and *Love Is Enough* have initials printed in blue.

F ORNAMENT FOR THE WELL AT THE WORLD'S END
One of the handsomest of the Kelmscott Press books is the edition of Morris's late romance, *The Well at the World's End*. It is set in two columns, with very graceful vine ornaments in the borders and between the columns. Morris's design for the last of these intercolumnar ornaments was drawn directly on a trial page of the book. The design as cut was not used on this page, 167, but makes its first appearance on page 322.

79

THE GOLDEN LEGEND
3 vol. 3 November 1892

A PAPER COPY [PML] (there were no vellum copies)
As has been mentioned, Morris purchased a copy of de Worde's 1527 edition of *The Golden Legend*, the English version of the most popular medieval collection of saints' lives, in the summer of 1890. This gave a focus to his typographic plans; he determined to print an edition of *The Golden Legend* in his new type which was now in process of being cut. In September, he made an agreement whereby Bernard Quaritch would publish the book (see below). *The Golden Legend* was preceded by six smaller Kelmscott Press books chiefly because a larger sheet of paper was required for it than that originally supplied by Batchelor. Caxton's original 1483 edition of *The Golden Legend* was used as copy text. The two

woodcuts by Burne-Jones were his first original work for the Kelmscott Press.

B DRAFT AGREEMENT WITH QUARITCH
    FOR THE GOLDEN LEGEND
   [Sanford L. Berger]

F. S. Ellis wrote to Quaritch on 4 September 1890, outlining his and Morris's requirements concerning *The Golden Legend*. On the 8th, Quaritch returned a draft agreement whose salient clause reads, 'Mr. Morris to have absolute and sole control over choice of paper, choice of type, size of the reprint, and selection of the printer.' In a covering letter, Quaritch referred to the recent success of the 'King Arthur' (i.e., Malory's *Morte d'Arthur*) published by David Nutt, and wrote, 'I ask from curiosity, will [Morris] make our G.L. uniform with Nutt's K.A.? Nutt printed a few copies of his book on Large Paper.' This makes for mildly ironic reading now, for Nutt's edition of Malory, though of considerable scholarly importance, is repulsively ugly and printed on execrable paper. It is apparent that at the time he made his agreement, Quaritch had no clear idea of Morris's intentions for *The Golden Legend*. The final agreement with Quaritch was signed on 11 September.

80

THE RECUYELL OF THE
HISTORYES OF TROYE
2 vol. 24 November 1892
[PML: gift of John M. Crawford, Jr.]

A PAPER COPY
Presentation inscription from Morris to Sydney Cockerell, 29 November 1892, with a note by Cockerell, 'This copy is made up of the sheets that were brought in from the Press for William Morris to inspect as each one was finished—when the books were finished he usually made me a present of the sheets.' Cockerell inserted in the volume a note from Morris's diary for February 1892, 'Recuelles began printing,' and Morris's autograph description of the book for Quaritch's catalogue, closing with 'surely this is well worth reading, if only as a piece of undiluted Mediaevalism.' The *Recuyell* contains the first use both of Troy and Chaucer type, the latter being used to print the Table of Chapters and glossary. It was issued only three weeks after *The Golden Legend*.

B MORRIS'S DESIGN FOR THE TITLE
This was Morris's second woodcut title, the first being done for *The Golden Legend*.
PLATE LXXXIV

81

THE STORY OF THE GLITTERING
PLAIN, 17 February 1894

A PAPER COPY [PML]
The Story of the Glittering Plain was the first book issued by the Kelmscott Press, in the small quarto format adopted for all of the first six books. The second Kelmscott Press edition was a much more elaborate volume, printed as a large quarto on Perch paper. It contains twenty-three woodcuts after designs by Walter Crane, making it second only to the Chaucer in amount of illustration. The illustrations are not, however, a great success. They were engraved by A. Leverett, whose rather fussy, ragged lines reproduce Crane's pen-work with great accuracy, but do not sort well with the sharp, clear typography.

B THREE WALTER CRANE DRAWINGS FOR THE STORY OF THE GLITTERING PLAIN

[PML: gift of John M. Crawford, Jr.]

The drawings represent the Sea-eagle striding out of the forest (p. 71), 3¼×4¼ in.; the Sea-eagle and Hallblithe before the king of the Glittering Plain (p. 77), 5½×4⅜ in.; and Hallblithe, the Sea-eagle, and the Sea-eagle's damsel (p. 98), 4⅜×4⅜ in.

PLATE LXXXV

82

PSALMI PENITENTIALES, 10 December 1894

A PAPER COPY [PML]

The *Psalmi Penitentiales* was one of two texts printed at the Kelmscott Press based on Morris's own manuscripts. The source of the psalms (in an English metrical version) was a fifteenth-century Book of Hours of the use of Gloucester, now PML M.99. This version is also found, with many textual variants, in more than a dozen other English manuscripts, one of which attributes them to Richard Maidstone (d. 1396), a Carmelite of Aylesford.

B MISCELLANEOUS PROOF SHEETS

[PML: gift of John M. Crawford, Jr.]

Bound with the proofs is Ellis's draft of the glossary printed at the end of the book. The glossary has been worked over heavily by Morris, with many deletions of words and definitions he thought superfluous. He was particularly severe on Ellis's habit of defining an archaic spelling by a paraphrase instead of the obvious modern equivalent. For Ellis's stef=rigid Morris wrote, 'stiff it is usually called'; and for schene=clear, 'sheen—I wonder since you come from *Sheen* hodie Richmond.' An accompanying letter by Morris apologizes for all the changes, 'but you told me to knock the glossary about; now I have done so.'

83

THE TALE OF BEOWULF, 2 February 1895

Following his numerous Icelandic translations, it was natural that, if Morris were to make any versions of Old English literature, he would choose *Beowulf*. His *Beowulf* translation, planned in 1892, was made in close collaboration with the Cambridge Anglo-Saxon scholar A. J. Wyatt, who was editing *Beowulf* for the Cambridge University Press, and with whom Morris read through the poem.

A WYATT'S TRANSLATION OF BEOWULF. 49 (of 60) leaves.

[PML: gift of John M. Crawford, Jr.]

Wyatt made a close, almost literal translation of *Beowulf* for Morris's guidance, including parenthetical explanations of the more obscure turns of phrase and kennings. Wyatt was paid £100 by the Press for his share in the translation.

B MORRIS'S TRANSLATION OF BEOWULF, EARLY DRAFT. 69 leaves. [PML]

This early draft contains many pencil revisions, and the first several hundred lines have been copied twice, with variants.

C MORRIS'S TRANSLATION OF BEOWULF, FINAL DRAFT. 105 leaves.

[PML: gift of John M. Crawford, Jr.]

This draft was printer's copy for the Kelmscott Press edition; it is a very handsome example of Morris's normal bookhand. He has marked locations of ornamental initials and red printing.

PLATE LXXXVI

D   KELMSCOTT PRESS EDITION, PAPER
     COPY
     [PML: gift of John M. Crawford, Jr.]
Emery Walker's copy, with his bookplate (printed at the Kelmscott Press after Morris's death); presentation inscription, 3 November 1932, from Walker to T. E. Shaw (Lawrence). At the time the presentation was made, Emery Walker was printing Lawrence's translation of the *Odyssey* in a limited edition.

PLATE LXXXVII

E   MORRIS'S DESIGN FOR THE WOODCUT
     TITLE [PML]
The design is drawn on a lightly inked proof of the woodcut border.

84

THE LIFE AND DEATH OF
JASON, 5 July 1895
[PML: gift of John M. Crawford, Jr.]

A   PAPER COPY
Presentation inscription from Morris to Sydney Cockerell, 8 July 1895, with a note by Cockerell, 'This copy consists of the sheets that were brought in from the Press for William Morris to inspect as each one was finished.'

B   FOUR DRAWINGS BY SIR EDWARD
     BURNE-JONES
Burne-Jones contributed the two illustrations for *Jason*. The one drawing of Jason with the golden fleece is very close to the frontispiece woodcut. The second woodcut, Medea in the presence of Circe, is represented by three drawings. The first, a rough sketch, has Circe sitting within a hall (as Morris described the scene). The second drawing places the women rather awkwardly against a mountain landscape; the third drawing, from which the woodcut was made, puts them in a walled garden. Each drawing $7 \times 4\frac{1}{4}$ in.

PLATES LXXXVIII, LXXXIX

C   SIGNED ORDER FORM FROM ELLEN
     TERRY, 9 JUNE 1895, FOR A
     COPY OF JASON
Sydney Cockerell must have preserved this as an interesting keepsake.

85

POEMS CHOSEN OUT OF THE
WORKS OF SAMUEL TAYLOR
COLERIDGE, 12 April 1896
[PML: gift of John M. Crawford, Jr.]

A   PAPER COPY
With Cockerell's note, 'This book consists of the sheets that were brought in for Morris's inspection as each one was printed.' It is bound in holland boards rather than the edition binding of limp vellum. The Coleridge, edited by F. S. Ellis, was the last of a Kelmscott Press series of English poets also including Herrick, Keats, Shelley, Tennyson, and Rossetti.

B   MORRIS'S DESIGN FOR THE WOODCUT
     TITLE
It is annotated by Morris, 'for Mr. Keates [the wood engraver] Dec. 17, 1895.'

PLATE XC

86

THE WELL AT THE WORLD'S
END, 4 June 1896

Concerning *The Well at the World's End* see also no. 78f. It was very long in printing, being first announced in December 1892. Cockerell attributes the delay to 'various reasons,' but the major reason surely was Morris's rejection of a series of illustrations made for the book by A. J. Gaskin, a young Birmingham artist. Gaskin first visited Kelmscott House in December 1892, with samples of his work, and he must have been given his commission soon after. Morris planned an extensively illustrated book, and Gaskin was paid £250 for his wood engravings. But in the event, none were used, and *The Well at the World's End* finally appeared with only four Burne-Jones illustrations, one at the head of each book.

A PROOFS AND CANCELLED SHEETS
[PML: gift of John M. Crawford, Jr.]
These proofs, some on proof paper, some on Kelmscott paper, and two on vellum, show several early stages in the planning and design of the book. Eleven woodcuts by Gaskin in three different sizes appear, and notes by Morris for illustrations show that more were planned. Though Morris did not favor them, Gaskin's illustrations are beautifully cut, with great clarity of line. Morris may have disliked the handling of the faces, which in some cases are so simply stylized as to have a cartoonlike appearance. One of Gaskin's large cuts is shown above Burne-Jones's reworking of the same scene.
PLATES XCI, XCII

B PUBLISHED EDITION, PAPER COPY [PML]
This copy was formerly Swinburne's, with a presentation inscription from Morris dated 9 June 1896. It includes a design by Morris for one of the borders.
PLATE XCII

Gaskin later made a handsome series of woodcuts for the Kelmscott Press edition of Spenser's *Shepheardes Calender*, reproduced there by electrotype (issued 14 October 1896).

87

LAUDES BEATAE MARIAE
VIRGINIS, 7 August 1896
[PML]

The text of the *Laudes* was edited anonymously by Sydney Cockerell. Its source was the early thirteenth-century manuscript that Morris referred to as the Nottingham Psalter, acquired by him at the Batemans' sale, 25 May and after, 1893 (now PML M.103). This Psalter has now been associated with Reading Abbey, whose dedication on 19 April is marked in the Calendar in gold as *dedicatio ecclesiae*. The *Laudes* have elsewhere been attributed to Stephen Langton, Archbishop of Canterbury; they are recorded in various other manuscripts. The *Laudes* was the first of two books printed at the Kelmscott Press in three colors.

88

THE STORY OF SIGURD THE
VOLSUNG, 25 February 1898
[PML]

Sydney Cockerell wrote that 'The two borders used in this book were almost the last that Mr. Morris designed. They were intended for an edition of The Hill of Venus, which was to have been written in prose by him, and illustrated by Sir E. Burne-Jones. The foliage was suggested by the ornament in two Psalters in the library at Kelmscott House.' One of these must have been the Windmill Psalter, Morris's last acquisition, whose folded leaves with

strongly serrated edges are closely similar to those in Morris's border (see no. 25).
PLATE XCIII

89

THE WAYZGOOSE MENUS
[PML: gift of John M. Crawford, Jr.]

Morris honored the ancient tradition of the wayzgoose, or annual dinner and outing of the shop, held by custom near Bartholomewtide (24 August). The workmen designed the menus for these outings, using their ingenuity to overcome the constriction of the few types available. Their results transmuted the Kelmscott Press materials into respectable imitations of conventional Victorian typographic fancy work.
PLATE XCIV

90

FINANCES OF THE KELMSCOTT PRESS
Ledger and cost book of the Press, 1893–96
[PML: gift of H. P. Kraus]

The present ledger book, which has never been thoroughly researched, is filled with information on the financial workings of the Kelmscott Press. One-half the ledger contains the cost accounts of the individual pieces of printing done at the Press, 1893–96; the other half contains a profit-and-loss account, summary of wages paid, rent, reckoning of depreciation of equipment, and other miscellaneous records. This is apparently a rationalized record, drawn up after Morris's death, rather than a daily record, for all the 1896 expenses are uniformly completed as of 3 October, the day Morris died. The page illustrated records a portion of the Chaucer expenses.
PLATE XCV

91

KELMSCOTT PRESS ADDRESS BOOK, 1897
[PML: gift of John M. Crawford, Jr.]

This small ledger book has a label in Cockerell's hand reading 'Kelmscott Press. Addresses to which announcements were sent 1897.' About 850 names are entered. A selection of subscribers may be listed to show some of the range of contemporary interest in the Press: Lord Amherst of Hackney, Rev. Stopford Brooke, A. J. Balfour, Arnold Bennett, Sir Walter Besant, Gordon Craig, Earl of Crawford, Leopold Delisle (Morris had sent complimentary copies of *The Golden Legend*, *Troye*, and *Reynard* to the Bibliothèque Nationale), Austin Dobson, Dean Farrar, F. J. Furnivall, Israel Gollancz, Kenneth Grahame, Cecil Harmsworth, W. Holman Hunt, M. R. James, W. P. Ker, Sidney Lee, George Meredith, A. Edward Newton (care of 'Grollier Club, 29 East 32nd Street, New York'), F. Yorke Powell, A. W. Pollard, Val Prinsep, Earl of Rosebery, Michael Sadler (father of the collector), George Bernard Shaw, Ellen Terry, G. O. Trevelyan, H. Yates Thompson, Gleeson White.

Chaucer

The Works of Geoffrey Chaucer was in all respects the masterpiece of the Kelmscott Press. This was the final collaboration of Morris and Burne-Jones, oldest and closest of friends, and it fully expressed their mutual conception of the ideal character of an ornamented book. Sydney Cockerell first heard Morris mention the Chaucer project in June 1891, when he was setting to work on the Troy type; in

November, as that type was nearing completion, a trial setting of lines from Chaucer was printed in it (no. 72c). The Troy type proved too large for a book of the size of Chaucer, and the smaller version known as Chaucer type was cut in 1892. The Chaucer was first announced to Kelmscott Press subscribers in December 1892, when it was estimated that the volume would contain 'about 60 designs' by Burne-Jones, who was already working on the illustrations at this time. Actual printing of the Chaucer did not begin until August 1894. As Burne-Jones progressed the end receded, for he found it necessary to increase the number of illustrations. In a notice to subscribers in November 1894, Morris announced (besides an increase in paper copies from 325 to 425) that 'there will now be upwards of seventy' illustrations by Burne-Jones. On a rough sheet probably made about this time, Burne-Jones estimated the number of illustrations at '72 all and he [i.e., Morris] won't have more.' Finally there were eighty-seven woodcuts, for the design of which Burne-Jones was paid £1500. In January 1895 a third Albion press was acquired, so that two presses could work off the Chaucer full time, leaving a press free for the other Kelmscott projects. For about three months in the summer of 1895 there was concern over a yellow stain which showed up behind some of the heavily inked borders, and several experts on ink were consulted. It was found eventually that sun-bleaching would remove the stain. This faint stain, caused presumably by the linseed oil base of the ink, can be found occasionally in other Kelmscott Press volumes. Burne-Jones's drawings were completed in December 1895, and the last wood engravings made from them were ready at the end of March 1896, at which time Morris completed his final design for the title. Five years after the planning first began, the Chaucer was issued to subscribers, in June 1896.

92

THE TEXT OF THE CHAUCER
Autograph draft, Morris to the Secretary to the Delegates of the Oxford University Press
[PML: gift of John M. Crawford, Jr.]

Morris's undated request for permission to use Prof. W. W. Skeat's Chaucer edition as base text for the Kelmscott production is interesting also for his description of the nature of the project: 'May I ask you to be so kind as to lay before the Delegates the request to allow me to make use of Professor Skeat's Chaucer under the following circumstances. I am publishing a Chaucer which is being printed at the Kelmscott Press which will be ornamented with picture-designs by Sir Edward Burne-Jones, and borders &c by myself: this book I hope to make a specially beautiful one as to typography & decoration and naturally wish to make the text as good as possible; I ask therefore to be allowed to avail myself of the corrections of errors which Professor Skeat's learning and acumen are making public, though we (Mr. F. S. Ellis the editor and myself) by no means intend to produce a literal reprint of his text. We should be pleased to make full acknowledgment of our obligations to Professor Skeat in the book itself. I should add that it is impossible that this book could come into competition with the text now publishing at the U. Press; as only 325 Copies will be issued at a high price (£20) and that it will have neither notes nor commentary. It is in-intended to be essentially a work of art.' The battered copy of Skeat's Chaucer used at the Press is now at Yale.

93

MORRIS'S DESIGNS FOR CHAUCER
[PML: gift of John M. Crawford, Jr.]

Chaucer was both the most heavily illustrated and the most heavily decorated of the Kelmscott Press volumes, and Morris was responsible for all the decoration. More than sixty separate designs of title, borders, frames, and large initials were created over a three-year period. Only Morris's original drawings for woodcuts are listed here, although John Crawford's gift to the Morgan Library includes also a large number of woodcut proofs of these designs.

A REJECTED BORDER DESIGN
Cockerell records that Morris began his first folio border on 1 February 1893, eighteen months before printing began, but that he became dissatisfied and did not complete it. The present drawing conforms to that description, and gives a good sense of the manner in which Morris worked. The border is drawn around an early Chaucer trial page consisting of many repeated lines from the 'Knight's Tale.'

PLATE XCVI

B FIRST PAGE OF TEXT
Morris worked at the decoration of the first page of text in February 1893. A proof of the letterpress was pulled, with a pasted-in impression of Burne-Jones's first woodcut. Morris added around this designs of the large border, the frame for the woodcut, the two-line head-title, and the initial WHAN, all of which were cut by W. H. Hooper on a single block. The setting of this proof is different from that of the finished Chaucer, and a different initial B is used here.

PLATE XCVII

C FOUR DESIGNS FOR FRAMES
Eighteen different frames were used for Burne-Jones's illustrations. Three of the designs here are for the frames used on pages 114-115, on page 139, and on pages 163 and 248. The fourth design, drawn around a proof of the woodcut for the 'Wife of Bath's Tale,' was cut but not finally used in the Chaucer; it was used on one of the London County Council certificates printed at the Press.

PLATE XCVIII

D EIGHT DESIGNS FOR LARGE INITIALS
Twenty-six large initial-words were used in the Chaucer. The original designs here are for THER, pages 88, 127, 132; IN, pages 112, 134, 160, 445; AMONG, page 137; FRO, page 138; AT, pages 152, 423; I HAVE, page 384; AFTER, page 421; and GLORY, page 426.

PLATE XCIX

E DESIGN FOR THE PRINTER'S MARK
A new 'Kelmscott' printer's mark was used in the Chaucer. Morris designed it on a proof of the last page, fitting it in among the text, border, woodcut illustration, and frame.

PLATE C

F DESIGN FOR THE TITLE PAGE
The title page was Morris's last design for the Chaucer. He began it in February 1896, after Burne-Jones confessed to some anxiety over the delay. The work on the title went slowly, being finished at the end of March. Some final retouching was done in May, and the title was printed as the final sheet through the press.

PLATE CI

94

SIZE ESTIMATE
8½ × 8 in.
[PML: gift of John M. Crawford, Jr.]

This casting-off estimate of the bulk of the Chaucer must have been made by Morris in July 1894, several weeks before printing began. On the 30th of that month he wrote Cockerell (in a letter given to the Morgan Library by John Crawford), 'having gone over the number of lines [in Chaucer] with Ellis, I find it will not make more than 600 pages, which will go into one volume.' Morris's estimate of 544 pages was very close. The completed Chaucer contains 553 pages of text, and some of the increase is surely due to Burne-Jones's decision to add more illustrations.

PLATE CII

95

SIR EDWARD BURNE-JONES,
ELEVEN DRAWINGS FOR
CHAUCER
[PML: gift of John M. Crawford, Jr.]

The automatic association of Burne-Jones's name with the Kelmscott Press depends almost entirely on his role in the Chaucer. Before Chaucer appeared, Burne-Jones had designed only eleven original woodcuts for the Press, scattered through six volumes (see nos. 79, 84, 86). The main group of Burne-Jones's drawings for Chaucer is in the Fitzwilliam Museum, which holds the eighty-five known finished drawings for the woodcuts, more than 150 preliminary and intermediate drawings, and a sketchbook of early rough studies for a number of the scenes. John Crawford's gift to the Morgan Library comprises a group of eleven preparatory drawings which the artist's son, Sir Philip Burne-Jones, presented to Sydney Cockerell in 1919. Two of the drawings were not used as woodcuts, and the others all show variants from the final designs. References are given to Duncan Robinson's *Companion volume* to the 1975 Basilisk Press facsimile of the Kelmscott Chaucer, which reproduced as plates the eighty-five finished drawings, and as figures in the text a number of preliminary sketches.

A THE KNYGHTES TALE, PAGE 9
   (Robinson pl. 2)
Emelye outside the dungeon. In the finished drawing she is holding a book. 5⅛ × 6¾ in.

B THE KNYGHTES TALE, PAGE 23
Emelye in the Temple of Diana. The finished drawing (Robinson pl. 5) illustrates an entirely different scene. A reworked photographic reproduction of the present sketch is in the Fitzwilliam Museum (Robinson fig. 23, and cf. fig. 22). 4⅝ × 6¼ in.

PLATE CIII

C THE KNYGHTES TALE, PAGE 24
   (Robinson, pl. 6)
The Temple of Mars. 4⅝ × 6¼ in.

D THE TALE OF THE MANNE OF LAWE,
   PAGE 43 (Robinson pl. 8)
Custance cast forth on the sea. In this drawing she sleeps in the boat; in the final version she is upright, gazing forlornly over the water. 5⅛ × 6¾ in.

PLATE CIV

E THE PRIORESSES TALE, PAGE 60
   (Robinson pl. 10)
The Virgin appears in a vision to the murdered Christian child, and 'Me thoughte she leyde a

greyn upon my tonge.' In the final version the Virgin has only one attendant. 5⅛×6¾ in.

PLATE CV

F THE TALE OF THE WIFE OF BATH,
    PAGE 114 (Robinson pl. 12)
The knight with his withered fiancée. This drawing is illustrated by Robinson (fig. 30). 5⅛×6¾ in.

G THE TALE OF THE CLERK OF
    OXENFORD, PAGE 136 (Robinson pl. 18)
Grisilde in her bare smock; cf. Robinson figure 12. In the final version she walks through a desert rather than a wooded landscape. 5⅛×6¾ in.

PLATE CVI

H THE ROMAUNT OF THE ROSE,
    PAGE 243 (Robinson pl. 30)
The poet's first view of the enclosed garden. 5⅛×6¾ in.

I THE PARLEMENT OF FOWLES
This drawing was not finally used. It illustrates a scene outside the temple of Venus described on page 317 of the Chaucer:
Aboute the temple daunceden alway
Wommen enow, of whiche somme ther were
Faire of hemself, and somme of hem were gay;
In kirtels, al disshevele, wente they there. . .
5⅛×6¾ in.

PLATE CVII

J TROILUS AND CRISEYDE, PAGE 470
    (Robinson pl. 75)
Chaucer with his muse Thesiphone, one of the Furies, views Troy. 4⅝×6¼ in.

K TROILUS AND CRISEYDE, PAGE 537
    (Robinson pl. 84)
Chaucer and Thesiphone foresee the burning of Troy. 4⅝×6¼ in.

PLATE CVIII

96

THE TRANSITION FROM DRAWING TO WOODCUT

A NOTE BY ROBERT CATTERSON-SMITH, 1917 [PML: gift of John M. Crawford, Jr.]

Burne-Jones's drawings could not with any efficiency be copied directly in wood, for they lack the regularized linear scheme which wood engravers like to follow. An intermediate photographic process was necessary, which is clearly explained by Catterson-Smith, the man responsible for its execution: 'Emery Walker made a very pale print of a photograph (a platino) from Sir E. B.-J.'s pencil drawing—the exact size of the drawing—I then stuck the print down on stout cardboard, and, in order to avoid the expansion of the paper I put the paste on the cardboard first and then applied the paper print very quickly, so quickly that it had no time to absorb moisture and so expand, then I immediately ran a hot smoothing iron over it which at once dried the paste. Next I gave the print a thin even wash of Chinese white with a little size in it. The result was to get rid of everything but the essential lines. Next I went over the pale lines with a very sharp pencil, copying and translating from the B.-J. drawing which was in front of me. The lines of shading were put in in pencil. These shadow lines were very difficult as they had to be translations of very grey pencil tones, and they often took a long time to fit into their spaces. When the pencil drawing was finished all trace of the photograph had disappeared. Next came the inking over, which was done with a fine round sable brush and very black Chinese ink which I bought in bottles. By putting a little size in the Chinese white, as above mentioned, the ink was not absorbed by the white and so remained jet black—otherwise it would have become grey. To get a good brush for the purpose was not easy. I spread the hairs fan-wise and look[ed] with a magnifying glass to see where there were any irregular hairs, if there were I cut them out with a razor and so got an even point. When difficulties arose in the treatment of passages I consulted B.-J.—sometimes work by Albert Durer was consulted to see how he had dealt with passages. Some of the drawings were done over several times. I found the best result came from using the brush in almost a vertical position, this with a well charged brush gave a "fat" even line. W.M. made a distinction between a "fat" and an "acid" line, an acid line was /, a fat ///. Finally E. Walker made a photograph on the wood block and Hooper cut it.'

B PLATINOTYPES OF THE BURNE-JONES DRAWINGS [John A. Saks]

The entire series of retouched platinotypes, made as described above, is in the possession of John A. Saks. The first five Burne-Jones drawings were prepared in this fashion by Fairfax Murray, who did not, however, work quickly enough for Morris's liking. In an attempt to speed matters, Morris had the engraver, W. H. Hooper, cut the sixth block directly from Burne-Jones's drawing, but the result was not satisfactory. For the remaining drawings, the process of photography and redrawing was again used, with Catterson-Smith taking Fairfax Murray's place.

97

PROOFS OF THE CHAUCER WOODCUTS
[PML: gift of John M. Crawford, Jr.]

A collection of proofs, most in multiple impressions, of eighty of the Chaucer woodcuts. Many of these are annotated and retouched by Burne-Jones, and some are dated and initialed by the cutter, W. H. Hooper. Generally, a progression of pulls seems to have been made, starting with thin proof paper, moving to thicker proof paper, then to a rag wove paper, and finally, after corrections, to Kelmscott paper.

A THE ROMAUNT OF THE ROSE, PAGE 250

Burne-Jones asks for more than a dozen small changes, most involving lightening of lines.

PLATE CIX

B THE ROMAUNT OF THE ROSE, PAGE 252

Burne-Jones instructs Hooper that 'this print has been touched,' i.e., with china white, and notes the points where retouching has been added.

PLATE CX

98

THE INTEGRATION OF PICTURE AND DESIGN
A.L.S., Edward Burne-Jones to Charles Eliot Norton, 8 December 1894
[PML: gift of John M. Crawford, Jr.]

Norton, professor of art history at Harvard University, was an old friend of Burne-Jones; he had seen the Chaucer in progress, and not liked what he saw. Burne-Jones's defense, though humorously expressed, is at the same time a precise statement of his and Morris's shared aesthetic of beauty in books: 'And so you don't like the Chaucer—that is very sad—for I am beside myself with delight over it. I am making the designs as much to fit the ornament & the printing as they are made to fit the little pictures—& I love to be snugly cased in the borders & buttressed up by the vast initials—& once or twice when I have no big letter under me, I feel tottery & weak; if you drag me out of my encasings it will be like tearing a statue out of its niche & putting it into a museum—indeed when the book is done, if we live to finish it, it will be a little like a pocket cathedral—so full of design. & I think Morris the greatest master of ornament in the world—& to have the highest taste in all things—& it is clear that you must come over & bathe in Jordan.'

99

PRINTING PRESS USED FOR THE KELMSCOTT CHAUCER

[J. Ben Lieberman]

This is the third of the full-size Albion presses used at the Kelmscott Press. It was purchased at the end of 1894 from its maker, Hopkinson & Cope, for £52 10s, and bears their serial number 6551, and date of manufacture, 1891. The press was bought to speed up the printing of the Chaucer; Emery Walker recalls that the additional iron bands were forged onto it as reinforcement, due to the heavy impression required for the Chaucer. This press was later used at various times by C. R. Ashbee, who purchased the entire Kelmscott Press plant except for type and blocks, F. W. Goudy, and Melbert Cary.

100

THE SPECIAL CHAUCER BINDING

The ordinary edition binding for the Chaucer was the familiar linen spine with holland boards. In a notice of February 1896, Morris announced plans for four alternative special bindings, all of his own design, two to be executed by the Doves Bindery, and two by Messrs. J. & J. Leighton, who had done the plain vellum bindings for most of the other Kelmscott Press books.

A A.L.S., T. J. COBDEN-SANDERSON TO
   WILLIAM MORRIS, 21 FEBRUARY 1896
[PML: gift of John M. Crawford, Jr.]
'My dear Morris
    The binding of "Chaucer."
    I undertake to bind the Kelmscott Chaucer for you in boards & pigskin, & to supply the materials for binding, but not the tools for the decoration, which are to be supplied by you and to be your own.
    I undertake to do this in either or both of two ways, as you may direct: to wit, either
    1. in whole white pigskin, or
    2. in half white pigskin,
and to charge you for the binding in whole pigskin £10 and for the binding in half pigskin £6.
            I am, dear Morris,
            affectionately yours,
            T. J. Cobden-Sanderson'

B PRELIMINARY DESIGN FOR THE FULL
   PIGSKIN BINDING
[PML: gift of John M. Crawford, Jr.]
This unfinished design was obviously meant to be carried out in pigskin, and must be an early version of the one binding design Morris was able to complete.
PLATE CXI

C THE WORKS OF GEOFFREY CHAUCER.
   20 JUNE 1896. [PML]
Due to his illness, the present binding is the only one of the four planned that Morris was able to complete. It is a remarkable reinter-

pretation of a fifteenth-century German binding style. The grape-vine border of the upper cover, it has been pointed out, is related to that of the title and facing page. The lower cover has an entirely different design. It is tooled in diagonal bands forming large lozenges, and within each lozenge, raised in relief, are stylized oak leaves. This design was derived from a binding in Morris's library, that of Ulrich Schreier of Salzburg, 1478, already mentioned. Schreier's binding is on a large folio Koberger Bible, very close in size to the Kelmscott Chaucer. Its lower cover in particular served to inspire the Chaucer binding: it too bears relief oak foliage with fillets, like those on the Chaucer, to suggest the leaf veins, and with a similarly centered rosette tool on each leaf. The combination of elements is characteristic of Austrian, and particularly Viennese, bindings of this period. The small flower tools in the center panel of the upper cover of the Chaucer can also be related to those on Schreier's binding. Many of Schreier's bindings combined cut-leather work and tooling, the cut-leather often having a criblée background. The small dots on the Chaucer binding's grape-vine border seem to have been inspired by this criblée effect. Morris's binding is nonetheless a work of art in its own right, the lower cover especially having a clarity and simplicity that excels its model. The numerous tools for the binding were engraved after Morris's designs, and the first two bindings to this pattern were executed by Douglas Cockerell and presented to Morris and Burne-Jones on 2 June 1896. By the end of 1897 forty-eight copies of the Chaucer had been bound thus, and an undetermined number were done at the Doves Bindery after that date. The binding of the present copy is dated 1898; it formerly belonged to Junius Spencer Morgan, nephew of Pierpont Morgan. 17 × 11 ¾ in.

PLATES CXII, CXIII

101

THE WORKS OF GEOFFREY CHAUCER
Kelmscott Press, 1896

A PAPER COPY
[PML: gift of John M. Crawford, Jr.]
This copy was given by Morris to Sydney Cockerell, his presentation inscription being dated 7 July 1896. Cockerell inserted in the volume a number of documents, including Morris's draft to the Oxford University Press (no. 92), a letter to Cockerell concerning sale of copies to Quaritch, Morris's estimate of the size of the volume (no. 94), his autograph draft of the colophon, and proof of the colophon, with corrections. On a back flyleaf Cockerell collected the signatures of seven principal contributors to the Chaucer. The copy is inserted in a chintz cover designed by Morris which Jane Morris kept on her own copy of the Chaucer, and which she later gave to Cockerell.

PLATE CXIV

B VELLUM COPY [PML]
Thirteen copies were printed on vellum. This copy bears a presentation inscription, written in a sadly shaky hand, from Morris to F. S. Ellis, dated 7 September 1896. Ellis later had the volume bound in oak boards by Douglas Cockerell.

PLATES

942 Liber moralitatum ** Lumen Anime dictus.
    1477. (Gunther Zeiner Augsburg).

943 Pii Secundi Pont: Max: Epistolæ.
    Mediolani impressit Antonius Zarotus 1481

944 Annronius de gestis Francorum &c.
    Paris Ascensius 1513.

945 ~~Albertus Magnus de Eucharistia~~
    Dionysii Cœlestis hierarchia. Paris per Henricum
    Stephanum 1515

946 Incipiunt duodecim quodlibeta beati Thome
    de acquino disputata. Sub anno domini
    millesimo CCLXXII. (This is written in red)
    Sine nota sed Romæ per Georgium Lauer c. 1470.

947 Guillelmi Caoursin Rhodiorum Vice Cancellarii
    obsidionis Rhodiæ Urbis descriptio
    Impressum Ulmæ p. Joannem Reger Anno dni
                                              MCCCXCVI.

948 Rationale Divinorum Officiorum.
    per ** Joannem Zeiner. Ulm 1475.

949 Incipit libellus de vita & Moribus philosophorum &
    poetarum.
    Sine nota. Antony Koberger? (Hain)

950 Boccatius de claris Mulieribus.
    Explicit Compendium Johannis Boccatii de Certaldo
    quod de preclaris mulieribus ac fama ppetuam
    edidit feliciter —
    S.n. sed Georg Husner Argent.

951 Caii Crispi Salustii de Lucii Catilinæ Conjuratione
    liber feliciter incipit. S.n. sed Martin Flach Argent.
    The whole book has the Jugurthine war also

books printed before 1500 a.d.

DONATUS in Terentium
Qui cupit obstrusam frugem gustasse Terenti
Donatum querat noscere grammaticum
Quem Vindelinus signis impressit ahenis
Vir bonus & claro preditus ingenio
sine anno, sed c. 1470
a large copy; but 7 leaves in MS. contemporary, or thereabout.

VAL. Catulli Veronensis poetæ doctissimi liber. P. Papini Statii Surculi Silvarum Liber
Impressum Parmæ per me Stephanū Corallū Anno Christi MCCCCLXXIII.

What's the probable date of the Speculum Salvationis

Augsburg. Gunther Zainer.

Dated books. 1471-1472.   Golden Legend.   140 cuts
              1472.   Ingold: Das goldene Spiel.   12 cuts
                          Eyb; ob einem Manne tzu nemen ein weib.   half border
              1477.   Cessolis. Schachbuch.   15 cuts

Undated.   ? 1475.   Rodericus Zamorensis: Spiegel des menschlichen lebens.
           c. 1473. Epistles & Gospels.   Full page cut of Christ & about 50 small square cuts.
           c. 1473. Die Bibel.
           1475.   Tuberinus: Die geschicht von Symon.   Interlaced initial I & 13 cuts.

           Æsop.
           Barlaam und Josaphat.
           Columna: Hystori von Troya.
           Belial.

This is (if not the first) yet one of the first of Gunther Zainer's woodcut books, and consequently one of the earliest of the Augsburg woodcut books: it is an early example of Zainer's 2nd gothic type. The woodcuts designed clearly by designer, are rude, but much character though they are not so decorative as some others of the Augsburg & Ulm books. The colouring of this copy is clearly original, and is systematic & good, the brownish lake especially is very good colour & contrasts well with the transparent yellow. The book lacks the added beauty of Zainer's large woodcut letters, which were probably not then done. The big Initial I very skilfully composed of simple strap-work is one of a few letters which (I believe) occur only in Gunther Zainer's books

(Note by William Morris)

Manuscripts.

1. Alexander de Villa Dei. Doctrinale. circa 1530. 3 10
2. Ambrosii Opuscula. Folio. 10
3. Antiphonarium sec. us. Rom. circa. c. 1380 1275 22 G Dunn
4. Antiphonarium. 12
5. Apocalypsis S. Johannis, cum comment. N. de Lyra. c. 1420. 40
6. Aretinus (Leon.) Historia Florentina. circa 1450.
7. Aristotelis, libri de virtutibus. circa 1450. 5
8. Athanasius contra gentiles. circa. 1465 15
9. Augustinus. Sermones super Psalmos. c. 1150 50 (Hodson)
10. " Manuale. xv saeculo 1462. im. 4to 3 Sem 99
11. Bartholomeus de Urbino. Dicta S. Ambrosii. 1458. Folio. 20
12. Biblia Sacra Latina. French. sm c. 1250. Im. Folio. Hodson Hornby 200
13. " " " 3 vols. folio. — Sem
14. " " " French, c.1250 Im. Folio. red mor Stratford Hodson Perrins 130
15. " " " French. c. 1290 120
16. " " " Italian c. 1260. 80 Boston sold

VI

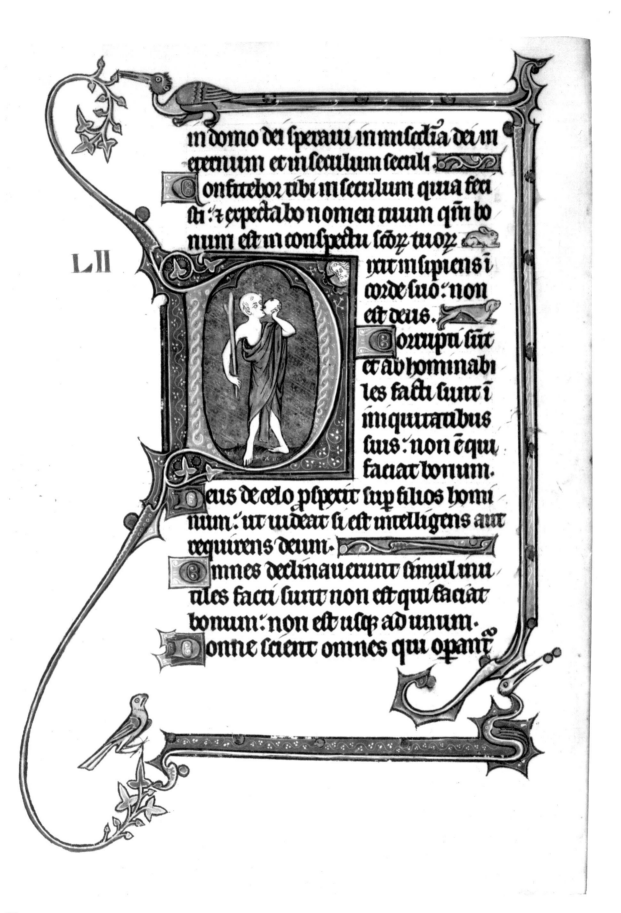
in domo dei speraui in mi sedia dei in
eternum et in seculum seculi.
Confitebor tibi in seculum quia fecisti: τ expectabo nomen tuum qm bonum est in conspectu scōp tuoψ.
[D]ixit insipiens in corde suo: non est deus.
Corrupti sūt et abhominabiles facti sunt in iniquitatibus suis: non ē qui faciat bonum.
Deus de celo pspexit sup filios hominum: ut uideat si est intelligens aut requirens deum.
Omnes declinauerunt simul inutiles facti sunt non est qui faciat bonum: non est usq̄ ad unum.
Nonne scient omnes qui op̄ant̄

VIII

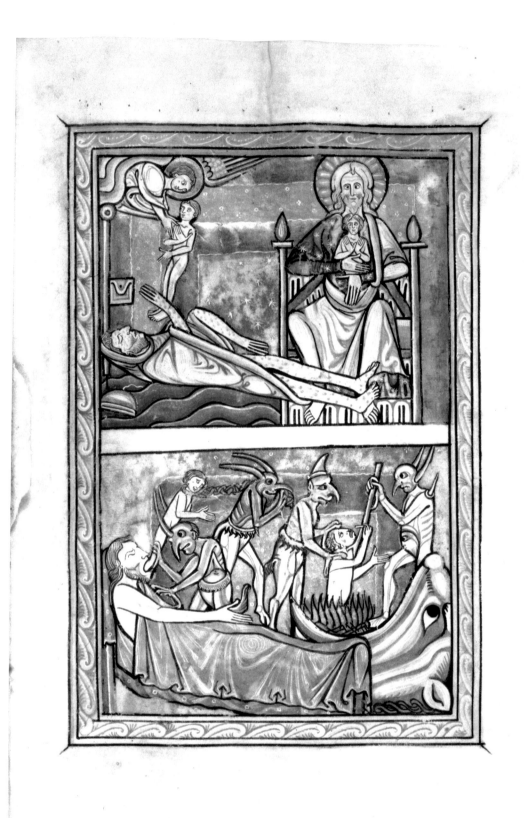

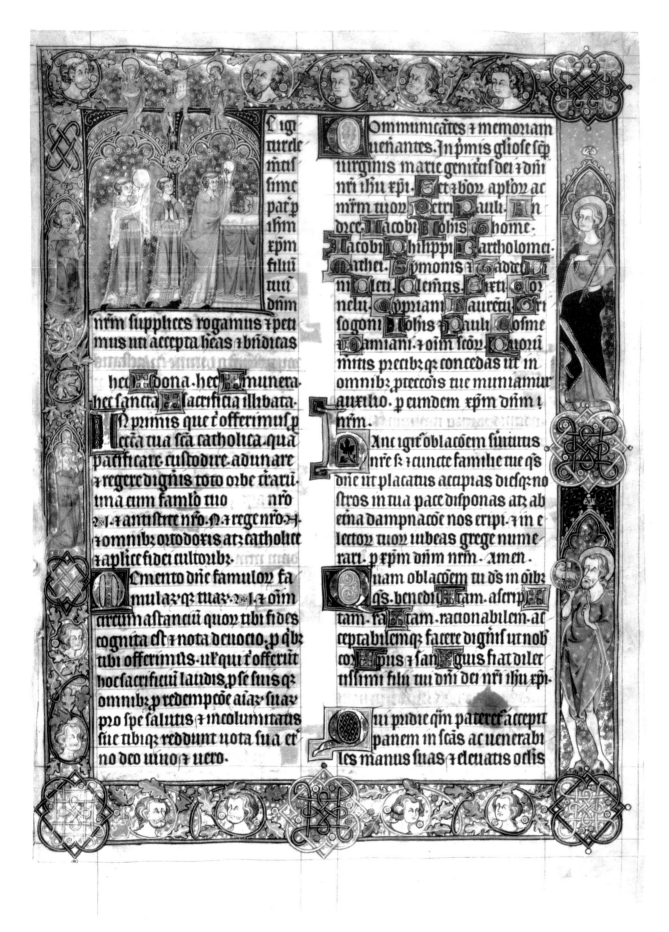

XII

GALA

Paulus apostolus non ab hominibus neque per hominem, sed per Ihesum Christum et deum patrem qui suscitavit eum a mortuis, et qui mecum sunt omnes fratres: ecclesiis Galathie. Gratia vobis et pax a deo patre nostro et domino Ihesu Christo, qui dedit semetipsum pro peccatis nostris, ut eriperet nos de presenti seculo nequam, secundum voluntatem dei et patris nostri: cui est gloria in secula seculorum, amen. Miror quod sic tam cito transferimini ab eo qui vos vocavit in gratiam Christi in aliud euangelium, quod non est aliud, nisi sunt aliqui qui vos conturbant, et volunt convertere euangelium Christi. Sed licet nos aut angelus de celo euangelizet vobis — preterquam quod euangelizavimus vobis, anathema sit. Sicut predixi et nunc iterum dico: si quis vobis euangelizaverit preter id quam quod accepistis, anathema sit. Modo enim hominibus suadeo an deo? An quero hominibus placere? Si adhuc hominibus placerem, Christi servus non essem. Notum enim vobis facio fratres euangelium quod euangelizatum est a me, quia non est secundum hominem. Neque enim ego ab homine accepi illud neque didici, sed per revelationem Ihesu Christi. Audistis enim conversationem meam aliquando in iudaismo, quoniam supra modum persequebar ecclesiam dei, et expugnabam illam, et proficiebam in iudaismo supra multos coetaneos meos in genere meo, habundancius emulator existens paternarum mearum traditionum. Cum autem placuit ei qui me segregavit ex utero matris mee, et vocavit per gratiam suam, ut revelaret filium suum in me, ut euangelizarem illum in gentibus: continuo non acquievi carni et sanguini: neque veni Ierosolimam ad antecessores meos apostolos: sed abii in Arabiam, et iterum reversus sum Damascum. Deinde post annos tres veni Ierosolimam videre Petrum, et mansi apud eum diebus quindecim. Alium autem apostolorum vidi neminem, nisi Iacobum fratrem domini. Que autem scribo vobis: ecce coram deo quia non mentior. Deinde veni in partes Sirie et Cilicie. Eram autem ignotus facie ecclesiis Iudee: que erant in Christo Ihesu. Tantum autem auditum habebant: quoniam qui persequebatur nos aliquando nunc euangelizat fidem quam aliquando expugnabat, et in me clarificabant deum.

c. Secundum.

Deinde post annos quatuordecim: iterum ascendi Ierosolimam cum Barnaba assumpto et Tito. Ascendi autem secundum revelationem: et contuli cum illis euangelium quod predico in gentibus. Seorsum autem hiis qui videbantur aliquid: ne forte in vacuum currerem aut cucurrissem. Sed neque Titus qui mecum erat cum esset gentilis: compulsus est circumcidi. Sed propter subintroductos falsos fratres qui subintroierunt explorare libertatem nostram, quam habemus in Christo Ihesu: ut nos in servitutem redigerent:

Que ret st obliuiscens ad destinatu: reddo biuu supne uocatiois. Et dns in euangliō dic. Dimitte mortuos sepelire mortuos. tu au uade. seqre me.

Tigris uocata ppt uolucrē fugā. ita ñ nominant pse. greci. & medi sagittā. Sit em bestia uarus distincta maclis. uirtute & uelocitate mirabilis. ex cui noīe flumē tigs appellat. qd his rapidissimū sit omniū fluuiox. has mag hircania gignit. Tygs ū ū uacuū rapte sobo

in consilio impiorum: & in uia pec
catorum non stetit: & in cathedra pe
stilentie non sedit.
Sed in lege domini uoluntas eius:
& in lege eius meditabit die ac nocte.
Et erit tanquam lignum qd plan
tatum est secus decursus aquarum:

XVIII

27

XIX

Das·xxx·capitel. Von d’ kunst des hirtenstabs· vñ irē nutz. Auch võ irē vngemach·ellend· vñ arbeyt

Die kunst der hirten· wöllent etlich· vnter dem ackerbaw begriffen sein. Etlich vnter der iägerei. Wie dem seye·so ist doch in sunderheyt davon zereden. Wann sy ist über nütz vnnd nottürftig den mangel der menschen züerfüllent. Dise ordnung zeleben ist

28

XXI

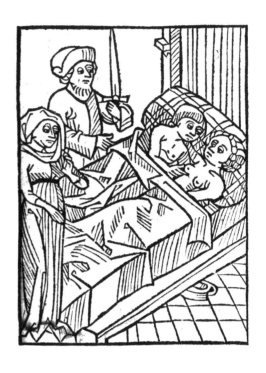

XXII

Cripturus igitur/quibus fulgoribus mulieres claruerint insignes/ a matre omniū sumpsisse exordium/ non apparebit indignum. Ea quippe vetutissima parens / vti prima sic magnificis fuit insignis splendoribus. Nam non in hac erumnosa miseriarum valle/in qua ad laboré ceteri mortales nascimū/pducta est/ nec eodem maleo aut incude etiā fabre fcta/ seu eiulans nascendi crimen deflens / aut inualida ceteroꝗ ritu venit in vitam/quinimo (qd̄ nemini vnqm alteri contigisse auditum est) cum iam ex limo terre reꝗ omnium faber optimus/ Adam manu cōpegisset ꝓpria/ & ex agro cui postea Damascenus inditum nomen est/ in orto deliciarum transtulisset/ cum in soporem soluisset placitum/ artificio sibi tm̄ cognito/ex dormientis late eduxit eandem sui comporem/& maturam viro/& loci amenitate atꝗ factoris letabundā intuitu/in mortalem/ & rerum dominam/ atꝗ vigilantis iam viri sociam/ & ab eodē Euam etiam nominatam. Quid maius/ quid splendidius potuit vnqm cōtigisse nascenti? Preterea hanc arbitrari possum? corporea formositate mirabilem quid enim dei digito fcm̄ est/ qd̄ cetera non excedat pulcritudine. Et qmuis formositas hec/annositate perita sit/ aut medio in etatis flore/ paruo egritudinis in pulsu lapsura/tn̄ quia inter ꝑcipuas dotes suas mulieres numerāt/& plurimum ex ea glorie(mortalium in discreto iudicio)iam consecute sunt/non superflue inter claritates earum /tanqm fulgor ꝑcipuus & apposita est/ & in sequentibus apponenda veniet. Hec insuper tam iure originis qm incolatus padisi ciuis fcta/& amicta splendore nobis incognito/dum vna cum viro loci delicijs

XXIV

XXVII

XXVIII

34

XXX

36

XXXI

XXXII

Ce chapitre est par maniere de prologue iusques ou il dist. Ce nest mie chose conuenable: ou le premier chapitre commence.

Incois que ie dye de linstitucion de lhōme ou il sera demonstre la naissance des deux citez tant cōme il touche et appartient aux creatures raisonnables mortelles/si cōme ou liure precedent il a este demonstre es anges/p lesquelles tant cōme nous pourons sera prouue: comment aux hōmes et aux anges/ compaignie ne soit mie dicte estre desconuenable ne mal seant/a ce que quatre citez/cest adire quatre compaignies ne soiēt mie dictes estre ordonnees Cestassauoir deux des anges: et deux des hōmes Mais qui plus est deux: cestassauoir Vne aux bons: laultre aux mauuais Non mie seulement aux anges: mais aux hommes.

Declaration de ce liure.

¶ Le prologue du compileur de ce present volume sur les fais du preup hector de troye.

EN ensuiuāt lordre comme cee ⁊ par moy promise a en suiuir a dame triumphe commanderesse de ceste euure ie apres la narracion des fais du noble et puissant alexandre ouquel eut tant de vaillances ⁊ conquestes, ie veuil commencer a escripre ses fais ⁊ grāt proesses du puissant et preup hector miroir de toute cheualerie ⁊ preus dommie. Et ne desplaise a aucun de ce que tant de louēge luy ay desia at

42

XXXV

XXXVI

XXXVII

XXXVIII

27

27

27

27

28

28

45

XXXIX

side went with their band, and great ill-will there was, and cross looks enow, but neither side set on other, but the men of Broadwick were the most at the market. Snorri the Godi rode in the evening to Temple-garth, whereas Biorn dwelt then, and his son Guest, the father of Ref of Temple-garth. The folk of Biorn the champion of the Broadwickers bade Arnbiorn to ride after those of Snorri the Godi, but Arnbiorn would not have it so, but said, that each should have what he had got; those of Snorri rode home the next day, and the sons of Thorbrand were worse content with their lot than heretofore.

Chap: XLIII

Now goodman Thorbrand had a thrall, who was called Egill the Strong, the biggest and strongest of men, and he thought his life ill, in that he was no free man, and would oft pray Thorbrand and his sons to give him his freedom, and offered to do therefor any such work as he might. So one evening as Egill went with his sheep in Burg-dale, as the evening grew late, he saw an erne fly from the west over the firth; now a great bear-hound was with Egill, and lo the erne stooped on the hound, and took him up in her claws, and flew back west over the firth unto Thorolf's cairn and vanished there, under the mountain, and great tidings this seemed to Thorbrand to forebode.

Now it was the wont of the Broadwickers in autumn at the beginning of Winter-nights to have ball-play under the Shoulder, south of Knerr; and the place thereof was called afterwards the Play-medds, and men betook themselves thither from all the country-side, and great play-booths were made there, wherein men abode, and dwelt there a half-month or more: great strength of chosen men was there from all the country-side, and many folk and much sport; most of the young men were at the plays, except Thord Bligr, but he might not deal therein because of his too great eagerness, though he was not so strong that he might not play for that cause; so he sat on a stool and looked on the play. Those brothers withal, Biorn and Arnbiorn, were not deemed meet to play

Within the city of the Solymi
As in this place, for what is thine as this thou builded & forty blis
 for all things there are thine yea even as this—
Then the King rose and filled a cup with wine—
And said praise be unto things divine—
Yet ere I pour, how goes it with our folk
Did many die before they laid the yoke
On these proud necks. When will they come again
O King they said, though they fell not in vain
Yet many fell, but now upon the way
The others are methinks on the third day
They will be here and needs must they be slow
Because they bear along as on they go
Wains full of spoil, arms and all fair attire
And silver that three times has known the fire
And men and women of the Solymi
According to thy will to live or die—
What sayest thou about Bellerophon
The King said that this day for me has won
Is he alive yet? Then the man waxed pale
And said he liveth, and of small avail
I steel against him certes, on the wall
He stood alone, for backward did we fall
Before the fury of the Solymi
Because we deemed ourselves brought thither to die
And might not heart—, then it was as though
A sudden light about his head did grow
Amidst the darts and clamour and he turned
A face to us that with such glory burned
And those behind us drave us back again
And cried aloud to die there in the plain
 leave him and with such a wave
Of desperate war swept up, they scarce could save
Their inmost citadel from us that
Who at the first with mocks had bid us bide
Until their slaves had got their whips in hand
To drive us thence — Now as he spake at first
The King as one who hears folk tell the worst
And must not heed it hearkened but at last
The rich noise from the Golden goblet cast
And said O Jove we thank thee take thou this
In token of the greatness of our bliss
In earnest of the good things that shall have
Who thus our name and noble friends didst save

So spake he looking downward and his heart
In what his lips said had but little part
And he who erst had told him of the thing
Seemed fain to linger as if yet the King
Had something more to say who yet no word
Had more for him but with great many a Lord
Made merry praising still Bellerophon
As he was at his hand. Great things had won—
But in her bower the fair Philonoe
Grew like the Queen of all delight to be
When she heard all who heretofore could bear
No voice among her women folk to hear
No sight to see but hateful was her state
Her very beauty but a thing to hate

This is of small avail
 it
though
turned heard
drave us
drave us back a
drave
96
though the
though scarce
wave though
desparate war sw
desparate war was
despa des
The King as one who hears
The

HERE BEGINS THE STORY OF THE DWELLERS AT EYR

CHAP: I.

Of Ketil Flatnose and his kin and how he won the South isles.

KETIL Flatnose was a mighty lord in Norway; he was the son of Biorn Bunu, the son of Duke Grim of Sogn; Ketil Flatnose was wedded, and had to wife Yngvild, the daughter of Duke Ketil Ram of Raumarik; Biorn and Helgi were their sons' names, but their daughters were these; Aud the Deeply-wealthy, Thorun the Horned, and Jorun the Manly-minded: Biorn the son of Ketil Flatnose was fostered east in Jamtaland, with that earl, who is called Kiallak, whose daughter was Giaflaug.

This was in the days when Harald Fairhair came to the rule of Norway, and before the face of that trouble many noble men fled from the land; some east over the Keel-ridge, some west over the sea, and some withal kept in the South-isles in the winter, and in summer

ERE BEGINS THE TALE AND TELLS of a man named Sigi, who was called of men the son of Odin; another man withal is told of, hight Skadi, a great man, and mighty of his hands, yet was Sigi of more might, and higher of kin, according to the speech of men of that time. Now Skadi had a thrall, with whom the story deals somewhat, Bredi was he hight, and was even named after that work that he had to do; in prowess and might of hand he was equal to men of more account, yea and better than some thereof.

Now it is to be told, that on a time Sigi fared to the hunting of the deer, and the thrall with him; day-long they hunted till the evening, and when they gathered their prey together at night-fall, lo greater and more was that which Bredi had slain, than was Sigi's prey: which thing misliked him much, and he said that great wonder it was, that a very thrall should out-do him in the hunting of deer; so he fell on him, and slew him, and buried the body of him thereafter in the snow. Then he went home in the evening, and said that Bredi had ridden away from him into the wild-wood: Soon was he out of my sight, says

THE TWO SIDES OF THE RIVER

THE YOUTHS

O winter, O white winter, wert thou gone,
No more within the wilds were I alone
Leaping with bent bow over stock and stone!

No more alone my love the lamp should burn,
Watching the weary spindle twist and turn,
Or o'er the web hold back her tears and yearn:
O winter, O white winter, wert thou gone!

THE MAIDENS

Sweet thoughts fly swiftlier than the drifting snow,
And with the twisting thread sweet longings grow,
And o'er the web sweet pictures come and go
For no white winter are we long alone.

HERE BEGINNETH THE STORY OF THE DWELLERS AT EYR: AND THIS FIRST CHAPTER TELLETH OF KETIL FLATNEB: AND OF HOW HE WON THE SOUTH ISLES

KETIL FLATNEB was a mighty lord in Norway: he was the son of Biorn Buna, the son of Duke Grim of Sogn: Ketil Flatneb was wedded, and his wife was Yngvild, the daughter of Duke Ketil Ram of Raumarike: Biorn and Helgi were their sons; and their daughters Aud the Deeply-wealthy, Thorun the Horned, and Jorun the Manly-minded: Biorn the son of Ketil Flatneb was fostered east in Jamtaland at an earl's called Kiallak, whose daughter was Giaflaug.

This was in the days when Harald Fairhair came to the rule of Norway; and before the face of that trouble many noble men fled away from the land, some east over the Keel-ridge, some west over the sea: and some withal abode in the South Isles in winter, and in

❦ Ogmund comes to Midfirth: those brethren told of:

Chap. II
Of Ogmund's settling in Iceland.

IN those days died king Harald Fairhair, and Eric Bloodaxe took the realm: but Ogmund had no friends in Eric and Gunnhild: so he arrayed his ship for Iceland. Ogmund and Helga had a son named Frodi; but whenas the ship was well-nigh ready Helga fell sick and died, and Frodi her son also. Then they put to sea, and Ogmund cast overboard the pillars of his high-seat.
So they made Midfirth, whereas the pillars had gone before them, and cast anchor there: in that time was Midfirth Skeggi ruler over those parts: he rowed out to them, and bade them come into the firth, and choose land there: so Ogmund took that bidding, and measured out the ground-plan of his house: now it was trowed in those days that if the measures met after often trying then would the goodman's fortune abide steady; but if the measure fell short so would his fortune also minish, but would grow greater if any were over of the measure: now here, the thing being tried thrice over, still would the measure meet. So there Ogmund built a house upon the knoll, and dwelt there thenceforward: he had to wife Dalla daughter of Onund Sioni, and their sons were Thorgils and Kormak: Kormak was black-haired, and his hair swept over his brow: he was pale of skin, and somewhat like his mother; a big man and a strong, and exceeding eager of temper. Thorgils was a silent man and a modest. But when those brethren were come to man's estate Ogmund died, and Dalla kept the household with her sons; and Thorgils ruled the stead with the furtherance of Midfirth Skeggi.

Chap. III
How Kormak first had sight of Steingerd.

THERE was a man called Thorkel, who dwelt at Tongue; he was wedded, and had a daughter named Steingerd who was now fostered at Gnupsdale.
Now one day in autumn a whale came ashore at Waterness which was the due of those brethren, the sons of Dalla; so Thorgils asked Kormak whether he had rather go to the fell-side or the whale, and he chose to go to the fell with the house-carles.
There was a man named Tosti, a foreman of theirs, who had heed of the sheep-fetching; with him went Kormak, and they fared on till they came to Gnupsdale, whereat they

THE STORY OF FRITHIOF THE BOLD

"A goodly gold ring hast thou."
"Yea in good sooth," said he.
Thereafter went those brethren to their own home, and greater grew their envy of Frithiof.
A little after grew Frithiof heavy of mood, and Biorn his foster-brother asked him why he fared so: he said that he had it in his mind to woo Ingibiorg: "For though I be named by a lesser name than those brethren, yet am I not fashioned lesser."
"Even so let us do then," quoth Biorn.
So Frithiof fared with certain men unto those brethren; and the kings were sitting on mound, when Frithiof greeted them well, and then set forth his wooing, & prayed for their sister, Ingibiorg the daughter of Beli.
The kings said: "Not overwise is this thine asking, whereas thou would have us give her to one who lacketh dignity; wherefore we refuse thee this utterly."
Said Frithiof: "Then is my errand soon sped; but in return never will I give help to you henceforward, yea though ye need it neversomuch."
They said they heeded it nought: so Frithiof went home, and was joyous once again.

CHAP. III.

THERE was a king named Ring, who ruled over Ringrealm, which also was in Norway; a mighty county-king he was, & a great man, but come by now into his latter days.
Now he spake to his men; "Lo I have heard that the sons of king Beli have brought to nought their friendship with Frithiof, who is the noblest of men; wherefore will I send men to these kings, and bid them choose whether they will submit them to me, and pay me scat, or else that I fall on them, and all things then shall lie ready to my hand, for they have neither might nor wisdom to withstand me; yet great fame were it to my old age to overcome them."
After that fared the messengers of king Ring, and found those brethren, Helgi and Halfdan, in Sogn, and spake to them thus: "King Ring sends bidding to you to send him scat, or else will he war against your realm."
They answered, and said, that they would not learn in their days of youth what they would be loth to know in their old age, even to serve king Ring with shame: "Nay, now shall we draw together all the folk that we may."
Even so they did; but now, when they beheld their force, that it was but little, they sent Hilding their fosterer to Frithiof to bid him come

THE STORY OF HALFDAN THE BLACK

Chap: 1 Halfdan fights with the Kings Gandalf and Sigtrygg.

HALFDAN was one winter old when his father died: Asa his mother went forthwith west to Agdir, and set herself down in the realm of her father Harald: there waxed Halfdan, and was big and strong in his early days, and was black of hair, wherefore was he called Halfdan the Black: when he was eighteen winters old he took the kingdom in Agdir; thereafter he went into Westfold and shared the realm there with King Olaf his brother, as is aforewrit.

That same autumn he went with an host against King Gandalf, and many battles they had together, and had the victory in turn; but in the end they made peace on such terms that Halfdan was to have the half of Vingulmark which his father Gudrod had owned. Thereafter fared King Halfdan up into Raumarik, and laid it under him, whereof heard the King Sigtrygg son of Eystein, who as then abode up in

WHEN Haroon al Rasheed had put to death his favourite Jaffier, and taken to him the goods and chattels of the whole family of the Bermukkies, the kin of Jaffier, he gave command that no man should should say a word in praise of their goodness under pain of heavy punishment. This command notwithstanding a grave and ancient man day by day would go to the perished wezeer's palace-stead, for the house had been razed to the ground. There would he sit down on a heap of rubbish, and lament bitterly, and set forth unto the passers-by the greatness and goodness of that unhappy family. Now when Rasheed was told of this he was very angry, and bade bring the man before

perhaps a pointed hand would be best on the sort
of quickness though at present it dont seem to
matter much: the ~~it~~ it was good. and the thick
pen if I could get it to go always

let us try some hands now this is not right
this is not right good but rather shaky.
tis somewhat of a puzzle to know how to set to
work about it: tis between pointed and round
a good piece of vellum ho   too broad

His burning eyeballs toward the town, fierce hear:

A good piece of work is not to be done with such
a very broad nibbed pen upon vellum with only
common ink, but I can write fast enough upon
it however: I wonder how it would do upon the
parchment: I dont like to write a nice MS on
common paper anyhow: the parchment was
no good it all ran together and would scacely
mark at all.
modo de temperare le penne
Con le uarie Sorti de littere ordinato per
   Ludouico Vicentino In Roma nel anno
no good at all: it all ran together and would

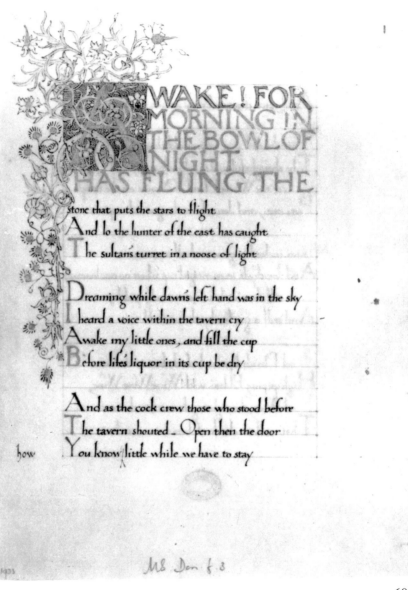

AWAKE! FOR MORNING IN THE BOWL OF NIGHT HAS FLUNG THE stone that puts the stars to flight
And lo the hunter of the east has caught
The sultan's turret in a noose of light

Dreaming while dawn's left hand was in the sky
I heard a voice within the tavern cry
Awake my little ones, and fill the cup
Before life's liquor in its cup be dry

And as the cock crew those who stood before
The tavern shouted — Open then the door
how You know little while we have to stay

the naming of certain men of Icefirth

THE STORY OF HAWARD THE HALT

Chap: 1 Of the Icefirthers

HERE beginneth this story, and telleth of a man named Thorbiorn the son of Thiodrek, who dwelt in Icefirth at a house called Bathstead, and had the Priesthood over Icefirth; he was a man of great kin and a mighty chief, but the most unjust of men, neither was there any in Icefirth of might to avail to gainsay him: he would take the daughters of men or their kinswomen, and handfast them awhile, and then send them home again: from some he took their goods, and other some he drave away from their lands: he had taken a woman, Sigrid by name, young and high born to be over his household; great wealth she had, which he would have on hand, and not put out to usury whiles she was with him

A man named Haward dwelt at the stead of Bluemire: he was of great kin, but come by now unto his latter days; in his earlier life he had been a great viking, and the best of champions; but in a certain fight he had gotten many sore hurts, and amongst them one

P. VIRGILII MARONIS

rursum in secessu longo sub rupe cavata
arboribus clausi circum atque horrentibus umbris
instruimus mensas, arisque reponimus ignem
rursum ex diverso coeli caecisque latebris
turba sonans praedam pedibus circumvolat uncis
polluit ore dapes. sociis tunc arma capessant
edico et dira bellum cum gente gerendum
haud secus ac jussi faciunt, tectosque per herbam
disponunt enses et scuta latentia condunt
ergo ubi delapsae sonitum per curva dedere
litora, dat signum specula Misenus ab alta
ære cavo: invadunt socii et nova proelia tentant
obscoenas pelagi ferro foedare volucres
sed neque vim plumis ullam nec vulnera tergo
accipiunt, celerique fuga sub sidera lapsae
semessam praedam et vestigia foeda relinquunt
una in praecelsa consedit rupe Celaeno
infelix vates, rumpitque hanc pectore vocem
Bellum etiam pro caede boum stratisque juvencis
Laomedontiadae, bellumne inferre paratis
et patrio Harpyias insontes pellere regno
accipite ergo animis atque haec mea figite dicta
quae Phoebo pater omnipotens, mihi Phoebus Apollo
praedixit, vobis Furiarum ego maxima pando
Italiam cursu petitis, ventisque vocatis
ibitis Italiam, portusque intrare licebit
sed non ante datam cingetis moenibus urbem
quam vos dira fames nostraeque injuria caedis

THE son of Swegdir was Vanland, and he took the realm after his father and ruled over the Wealth of Upsala; he was a great warrior and fought about the World. On a winter-tide he abode in Finland with Snow the Old, and wedded his Daughter Drift: but in the spring-tide he went his ways, and Drift was left behind: but he promised to come back in three winters' space; yet came he not in ten winters. Then sent Drift after Huld the witch-wife, and sent Visbur the son of Vanland into Sweden. Drift dealt with Huld the witch-wife in such wise that she was to draw King Vanland into Finland by magic, or slay him else. Now when the spells were sped Vanland was at Upsala, and straight he grew fain of faring to Finland; but his friends and councillors forbade it him, saying that the magic of the Fins was busy in his desire: then he became heavy with slumber, and laid himself down to sleep; but when he had slept a short space, he

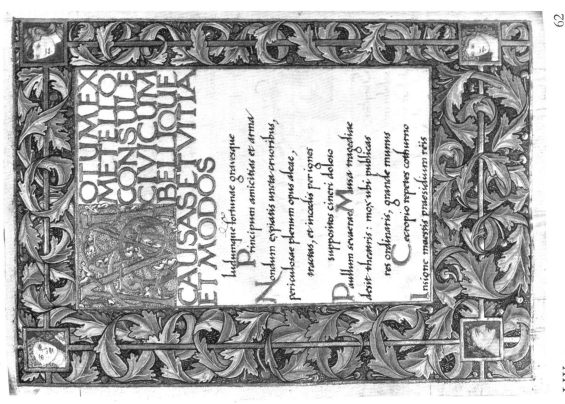

LIV-d

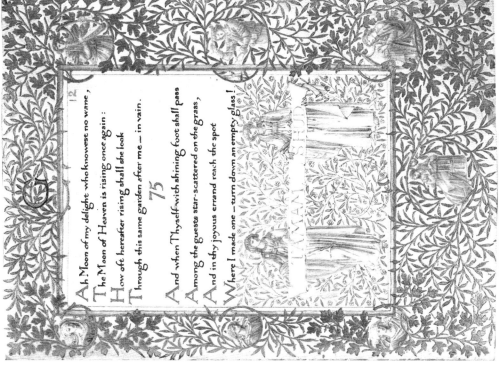

LIV-c

P. VIRGILII MARONIS

conserimus multos anatum demittimus rco
diffugiunt alii ad naves et litora cursu
fida petunt; pars ingentem formidine turpi
scandunt rursus equum et nota conduntur in alvo
EU mihi invitis fas quenquam fidere divis
ecce trahebatur passis riameia virgo
crinibus a templo assandra adytisque inervae
ad coelum tendens ardensia lumina frustra
lumina, nam teneras arcebant vincula palmas
non tulit hanc speciem furiata mente orœbus
et sese medium injecit periturus in agmen
consequimur cuncti et densis incurrimus armis
hic primum ex alto delubri culmine telis
nostrorum obruimur oriturque miserima cædes
armorum facie et raiarum errore jubarum
tum anai gemitu atque ereptæ virginis ira
undique collecti invadunt, acerrimus jax
et gemini tridæ olopumque exercitus omnis
adversi rupto ceu quondam turbine venti
confligunt ephyrusque otusque et lætus ois
rus equis; stridunt silvæ sævitque tridenti
spumeus atque imo ercus ciet æquora fundo
illi etiam si quos obscura nocte per umbram
fudimus insidiis totaque agitavimus urbe
apparent; primi clipeos mentitaque tela
agnoscunt atque ora sono discordia signant
ILICET obruimur numero primusque orœbus
enclei dextra divæ armipotentis ad aram

44

64A

64B

LV

The Story of Cupid and Psyche.

The Oracle.

O FATHER of a most unhappy maid, [know
O King whom all the world henceforth shall
As wretched amongst wretches, be afraid
To ask the Gods thy misery to show;
But if thou needs must have it, to thy woe,
Take back thy gifts to feast thine eyes upon
When thine own flesh and blood some beast hath won.

For hear thy doom, a rugged rock there is
Set back a league from thine own palace fair,
There leave the maid that she may wait the kiss
Of that fell monster that doth harbour there;
This is the mate for whom her yellow hair
And tender limbs have been so fashioned,
This is the pillow for her lovely head.

O what an evil from thy loins shall spring,
For all the world this monster overturns;
He is the bane of every mortal thing,
Nor any pity from the Gods he learns,
A fire goeth from his mouth that burns
Worse than the flame of Phlegethon the red;
To such a husband shall thy maid be wed.

And if thou sparest now to do this thing
I will destroy thee and thy land also,
And of dead corpses shalt thou be the king,
And stumbling through the dark land shalt thou go
Howling for second death to end thy woe.
Live therefore as thou may'st and do my will,
And be a king whom men may envy still.

The grief of all men.
What man was there who shuddered not for grief
At hearing this? Psyche, shrunk like the leaf
The autumn frost first touches on the tree,
Stared round about with eyes that could not see,
And muttered sounds from lips that said no word,
And still within her ears the sentence heard
When all was said, and silence fell on all
'Twixt marble column and fair gold-hung wall.

The lamentation of the King
Then spoke the King, bowed down with misery,
"What help is there, together let us die,
Or else, together fleeing from this land,
From town to town go wandering hand in hand
Thou and I, daughter, till all men forget
That ever in a throne I have been set,
And then, when houseless and disconsolate
We ask an alms before some city gate,
The Gods perchance a little gift may give,
And suffer thee and me like beasts to live."

But answered Psyche through her bitter tears, *and syche.*
"Alas, O Father, I have known these years
That with some woe the Gods have dowered me
As well as riches, and this vain beauty,
And ill it is against the Gods to strive.
O Father, live, for those that are alive
May still be happy: will it profit me
To live some months, and ere I die to see
You dying, and all folk that love me well,
And then at last myself go down to Hell,
Cursed of all men? Therefore make haste to lead
My body to the rock, and therewith feed
This monster, that the Gods so love or fear
They satiate him with a life held dear
Of many, and once called the world's delight.
That surely shall have ending on this night.
And yet, ah God, if I had been some maid
Toiling all day, and in the evening laid
Asleep on rushes, had I only died
Before this sweet life I had fully tried,
When on my mother's breasts I hung sucking,
And had no sense, or joyed in anything!"

And therewith she arose and gat away,
And in her chamber mowning long she lay
Thinking of all the days that might have been,
And how that she was born to be a Queen,
The prize of some great conqueror of renown,
The joy of many a country and fair town;
The high desire of every prince or lord,
She who could fright with careless smile or word,
The hearts of heroes fearless in the war;
The glory of the world, the leading-star
Unto all honour and all earthly fame,—
Round goes the wheel, and death and deadly shame
Is now her lot, while yet of her men sing,
Unwitting that the Gods have done this thing.
Long time she lay, the while the sunbeams moved

The Story of Cupid and Psyche.

Argument.

Psyche was a King's daughter, whose beauty made all people forget Venus, wherefore the Goddess hated her, and would fain have destroyed her: nevertheless she became the bride of Love, but her sisters gave her such evil counsel that he was wrath with her, and left her: whereon, she, having first revenged herself of her sisters, wandered through the world seeking him, and so doing, fell into the hands of Venus, who tormented her, and set her fearful tasks to accomplish; but the Gods and all nature helped her, so that at last she was reunited to love, forgiven by Venus, and made immortal by the Father of Gods and Men.

Part I.

N the Greek land of old there was
    a King   [thing,
Happy in battle, rich in every
But chiefly that he had a young
    daughter   [in her,
Who was so fair all men rejoiced

Psyche and her beauty.

So fair that strangers, landed from the sea,
Beholding her, thought verily that she
Was Venus visible to mortal eyes,
Fresh come from Cyprus for a world's surprise.
She was so beautiful, that had she stood
On windy Ida by the oaken-wood,
And bared her body to that shepherd's gaze,
Troy might have stood till now with happy days,
And those three fairest all have gone away
And left her with the apple on that day.

And Psyche is her name in stories old
As even by our fathers we were told;
All this saw Venus from her golden throne,
And knew that she no longer was alone

Psyche hated of Venus.

For beauty, but, if even for a while
This damsel matched her God-enticing smile:
Wherefore she wrought in such a wife, that she,
If honoured as a Goddess, certainly,

Was dreaded as a Goddess none the less,
And pined away long time in loneliness.

Two sisters had she, and men called them fair—
But as king's daughters might be anywhere,—
And these to lords of great name and estate
Were wedded, but at home must Psyche wait.
The sons of kings before her silver feet
Still bowed and sighed for her; in music sweet
The minstrels to all men still sung her praise,
While she must live a virgin all her days.

Her sisters wedded, but she a virgin.

So to Apollo's temple sent the King
To ask for aid and counsell in this thing,
And therewith sent he goodly gifts of price,
A silken veil wrought with a paradise,
Three golden bowls set round with many a gem,
Three silver cloaks, gold sewn in every hem,
And a fair ivory image of the God
That underfoot a golden serpent trod:
And when three lords with these were gone away
And must be gone now till the twentieth day,
Ill was the King at ease, and neither took
Joy in the chase, nor in the pictured book
The skilled Athenian limner had just wrought,
Or in the golden cloths from India brought.
At last the day came for those lords' return,

The King sends to the oracle.

LVIII

THE
EARTHLY PARADISE
A POEM.

BY
WILLIAM MORRIS,
AUTHOR OF THE LIFE AND DEATH OF JASON.

London : F. S. ELLIS, 33 *King Street, Covent Garden.*
MDCCCLXVIII.

[*All Rights reserved.*]

64ᴦ

LX

So then was Psyche taken to the hill,
And through the town the streets were void and still;
For in their houses all the people stayed,
Of that most mournful music sore afraid.
But on the way a marvel did they see,
For, passing by, where wrought of ivory,
There stood the Goddess of the flowery isle,
All folk could see the carven image smile.
But when anigh the hill's bare top they came,
Where Psyche must be left to meet her shame,
They set the litter down, and drew aside
The golden curtains from the wretched bride,
Who at their bidding rose and with them went
Afoot amidst her maids with head down-bent,
Until they came unto the drear rock's brow;
And there she stood apart, not weeping now,
But pale as privet blossom is in June.
There as the quivering flutes left off their tune,
In trembling arms the weeping, haggard King

LOVE IS ENOUGH.

THE MUSIC.

LOVE is enough: though the world be
 a-waning,
And the woods have no voice but the voice
 of complaining;
 Though the sky be too dark for dim
  eyes to discover
 The gold-cups and daisies fair blooming
  thereunder,
 Though the hills be held shadows, and the
  sea a dark wonder,
And this day draw a veil over all deeds passed over,
Yet their hands shall not tremble, their feet shall not falter;
The void shall not weary, the fear shall not alter
These lips and these eyes of the loved and the lover.

THE EMPEROR.

The spears flashed by me, and the ~~spears~~ *swords* swept round,
And in war's hopeless tangle was I bound,
But straw and stubble were the cold points found,
For still thy hands led down the weary way.

THE EMPRESS.

Through hall and street they led me as a queen,
They looked to see me proud and cold of mien,
I heeded not though all my tears were seen,
For still I dreamed of thee throughout the day.

LOVE IS ENOUGH.

THE EMPEROR.

Wild over bow and bulwark swept the sea
Unto the iron coast upon our lee,
Like painted cloth its fury was to me,
For still thy hands led down the weary way.

THE EMPRESS.

They spoke to me of war within the land,
They bade me sign defiance and command;
I heeded not though thy name left my hand,
For still I dreamed of thee throughout the day.

THE EMPEROR.

But now that I am come, and side by side
We go, and men cry gladly on the bride
And tremble at the image of my pride,
Where is thy hand to lead me down the way?

THE EMPRESS.

But now that thou art come, and heaven and earth
Are laughing in the fulness of their mirth,
A shame I knew not in my heart has birth—
—Draw me through dreams unto the end of day!

THE EMPEROR.

Behold, behold, how weak my heart is grown
Now all the heat of its desire is known!
Pearl beyond price I fear to call mine own,
Where is thy hand to lead me down the way?

THE EMPRESS.

Behold, behold, how little I may move!
Think in thy heart how terrible is Love;

LOVE IS ENOUGH.

THE MUSIC.

LOVE is enough: though the world be a-waning,
And the woods have no voice but the voice of complaining;
Though the sky be too dark for dim eyes to discover
The gold-cups and daisies fair blooming thereunder,
Though the hill be held shadows, and the sea a dark wonder,
And this day draw a veil over all deeds passed over,
Yet their hands shall not tremble, their feet shall not falter;
The void shall not weary, the fear shall not alter
These lips and these eyes of the loved and the lover.

THE EMPEROR.

The spears flashed by me, and the spears swept round,
And in war's hopeless tangle was I bound,
But straw and stubble were the cold points found,
For still thy hands led down the weary way.

THE EMPRESS.

Through hall and street they led me as a queen,
They looked to see me proud and cold of mien,
I heeded not though all my tears were seen,
For still I dreamed of thee throughout the day.

A TALE OF THE HOUSE OF THE WOLFINGS AND ALL THE KINDREDS OF THE MARK WRITTEN IN PROSE AND IN VERSE BY WILLIAM MORRIS.

WHILES IN THE EARLY WINTER EVE
WE PASS AMID THE GATHERING NIGHT,
SOME HOMESTEAD THAT WE HAD TO LEAVE
YEARS PAST; AND SEE ITS CANDLES BRIGHT
SHINE IN THE ROOM BESIDE THE DOOR
WHERE WE WERE MERRY YEARS AGONE
BUT NOW MUST NEVER ENTER MORE,
AS STILL THE DARK ROAD DRIVES US ON.
E'EN SO THE WORLD OF MEN MAY TURN
AT EVEN OF SOME HURRIED DAY
AND SEE THE ANCIENT GLIMMER BURN
ACROSS THE WASTE THAT HATH NO WAY;
THEN WITH THAT FAINT LIGHT IN ITS EYES
A WHILE I BID IT LINGER NEAR
AND NURSE IN WAVERING MEMORIES
THE BITTER-SWEET OF DAYS THAT WERE.

LONDON 1889: REEVES AND TURNER 196 STRAND.

THE ROOTS OF THE MOUNTAINS WHEREIN IS TOLD SOMEWHAT OF THE LIVES OF THE MEN OF BURG/ DALE THEIR FRIENDS THEIR NEIGHBOURS THEIR FOEMEN AND THEIR FELLOWS IN ARMS
BY WILLIAM MORRIS

WHILES CARRIED O'ER THE IRON ROAD,
WE HURRY BY SOME FAIR ABODE;
THE GARDEN BRIGHT AMIDST THE HAY,
THE YELLOW WAIN UPON THE WAY,
THE DINING MEN, THE WIND THAT SWEEPS
LIGHT LOCKS FROM OFF THE SUN-SWEET HEAPS—
THE GABLE GREY, THE HOARY ROOF,
HERE NOW—AND NOW SO FAR ALOOF.
HOW SORELY THEN WE LONG TO STAY
AND MIDST ITS SWEETNESS WEAR THE DAY,
AND 'NEATH ITS CHANGING SHADOWS SIT,
AND FEEL OURSELVES A PART OF IT.
SUCH REST, SUCH STAY, I STROVE TO WIN
WITH THESE SAME LEAVES THAT LIE HEREIN.

LONDON MDCCCXC: REEVES AND TURNER
CXCVI STRAND

Chapter ij. Of Thorstein's Dream.

One summer, it is said, a ship came from over the main into Gufaros. Bergfinn was he hight who was the master thereof, a Northman of kin, rich in goods, and somewhat stricken in years, and a wise man he was withal. Now, goodman Thorstein rode to the ship, as it was his wont mostly to rule the market, and this he did now. The Eastmen got housed, but Thorstein took the master to himself, for thither he prayed to go. Bergfinn was of few words throughout the winter, but Thorstein treated him well. The Eastman had great joy of dreams. One day in spring-tide Thorstein asked Bergfinn if he would ride with him up to Hawkfell, where at that time was the Thing-stead of the Burg-firthers: for Thorstein had been told that the walls of his booth had fallen in. The Eastman said he had good will to go, so that day they rode, some three together, from home, and the house-carles of Thorstein withal, till they came up under Hawkfell to a farmstead called Foxholes. There dwelt a man
of

tro faceuano grande infift
che erano gente dipregio.
ldo. Tucti quefti faccend
r lalunga moffeno efiore
aggiote fforzo italmanier
echati &foffi inpiu luog
teffi ne ufcire ne ētrare. I
lle man fra lemura della
fifaceuano apiftoia: Pap
decto nelpontificato p il

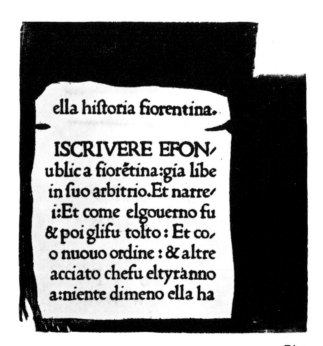

sua ruina ci pare uno
che lementi degli huomin
eglino possino rimunerar
do che epremii sono alle u
fficulta non e appresso ed
iamo quegli

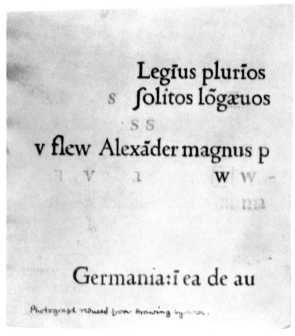

71E

71F 71F

in the fable of the fat baron or
the barbarian barber a fatal rain
of hot beer or herb tea on the
northern hill left one another to
tolerate the entire fare or to alter

Trial when only 11 letters were cut.

71G

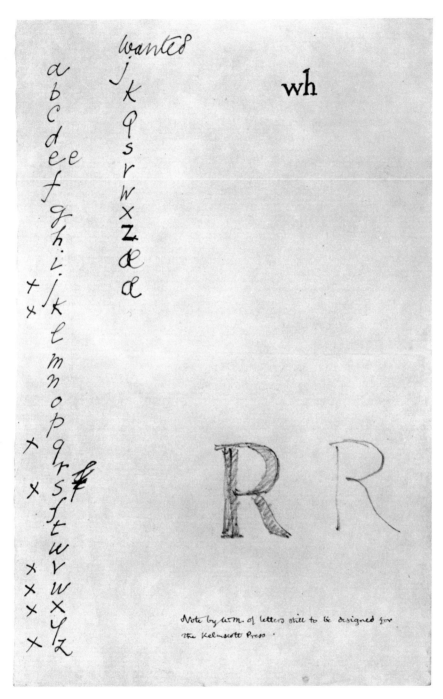

ten abcdefghijklmnop

illu d

esti abcdefghijklmnopqrstuvwxyz

erũ ABCDEFGHIJKLMNOP

m uit QRSTUVWXYZ

scip abcdefghijklmnopqurstvwxyz

uce abcdefghijklmnopqurstvwxyz

done

a a d b b c e f g m
h m i u u j k l n
o p q r s t u u
v y f f æ œ fl

Wher he schal have his love or fare amys,
Awayteth night and day on this miracle;
And whan he knew that ther was noon obstacle,
That voyded were these rokkes everichoon,
Doun to his maistres feet he fel anoon,
And sayd; 'I wrecched woful Aurilius,
Thanke you, lord, and my lady Venus,
That me han holpe fro my cares colde.'
And to the temple his way forth he hath holde,
Wher as he knew he schold his lady se.
And whan he saugh his tyme, anoon right he
With dredful hert and with ful humble cheere
Salued hath his owne lady deere.
'My soverayn lady,' quod this woful man,
'Whom I most drede, and love, as I best can,
And lothest were of al this world displese,
Nere it that I for you have such desese,
That I most deye her at youre foot anoon,
Nought wold I telle how me is wo bygoon,
But certes outher most I dye or pleyne;
Ye sleen me gulteles for verrey peyne.
But of my deth though that ye have no routhe;
Avyseth yow, or that ye breke your trouthe;
Repenteth yow for thilke God above,
Or ye me sleen, bycause that I you love.
For, Madame, wel ye woot what ye han hight;

and that othur knight hight
palamon nat fully quyk nes
fully deed they were but bye
here cooki armure and by her
g pilouis som and folk town
joy saturne constellacioung
heraudes knew hom wyl ing
bon jour mon fight dwyk m
dyk fgbqwx jpbz vpf

73

Unkingly, unhappy, he went his ways homeward.
  A Councillor. But by other ways yet had thy wisdom to travel;
How else did ye work for the winning him peace?
  Oliver. We bade gather the knights for the goodliest tilting,
There the ladies went lightly in glorious array;
In the old arms we armed him whose dints well he knew
That the night dew had dulled and the sea salt had sullied:
On the old roan yet sturdy we set him astride;
So he reached forth his hand to lay hold of the spear
Neither laughing nor frowning, as lightly his wont was
When the knights are awaiting the voice of the trumpet.
It awoke, and back beaten from barrier to barrier
Was caught up by knights' cry, by the cry of the king:
Such a cry as sets Mars in the Council-room winds
May awake with some noise that horn just sounded
And the bones of the world ringing to the
So it seemed to my heart, and mine eye,
As the spears met and splinters flew high o'er the field,
And I saw the king when his course was at swiftest,
His horse stayed on the bit, and he standing
Stiff and straight in stirrups, his spear held by the midmost,
His helm cast back, his teeth set hard together;
E'en as one might, who, riding to heaven, feels round him
The devils unseen: then he raised up the spear
As to cast it away, but therewith failed his fury,
  12

74

TYPES USED AT THE KELMSCOTT PRESS.

The Troy Type.

A B C D E F G H I J K L M N O P Q R S T U V W X Y Z

1 2 3 4 5 6 7 8 9 0

a b c d e f g h i j k l m n o p q r s t u v w x y z

æ œ & ff fi ffi fl ffl ! ? (' . , : ; - ,

The Chaucer Type.

A B C D E F G H I J K L M N O P Q R S T U V W X Y Z

1 2 3 4 5 6 7 8 9 0

a b c d e f g h i j k l m n o p q r s t u v w x y z

æ œ & ff fi ffi fl ffl ! ? (' . , : ; - ,

The Golden Type.

A B C D E F G H I J K L M N O P Q R S T U V W X Y Z Æ Œ

1 2 3 4 5 6 7 8 9 0

a b c d e f g h i j k l m n o p q r s t u v w x y z

æ œ & ff fi ffi fl ffl ! ? (' . , : ; ,

THE old man answered not a word, and he seemed to be asleep, and Hallblithe deemed that his cheeks were ruddier and his skin less wasted and wrinkled than aforetime. Then spake one of these women: Fear not, young man; he is well and will soon be better. Her voice was as sweet as a spring bird in the morning; she was white-skinned and dark-haired, and full sweetly fashioned; and she laughed on Hallblithe, but not mockingly; and her fellows also laughed as though it were strange for him to be there. Then they did on there shoon again, and with the carle laid their hands to the bed whereon the old man lay, and lifted him up, and bore him forth on to the grass, turning their faces towards the flowery wood aforesaid; and they went a little way and then laid him down again and rested; and so on little by little, till they had brought him to the edge of the wood, and still he seemed to be asleep. Then the damsel who had spoken before, she with the dark hair, said to allblithe: "Although we have gazed on thee as if in wonder, this it not because we did not look to meet thee, but because thou art so fair and goodly a man: so abide thou here till we come back to thee from out of the wood."

Therewith she stroked his hand, and with her

78A

I22 I19 I20 I18

78B

Note by Emery Walker
the Cattleson-Smiths delivered
Sydney Cockerell 23 Nov 1927

One row printed from wood; the other from electro types (done to ensure uniformity; but electros, if it was without arbitrary [?])

LXXXI

LXXXII

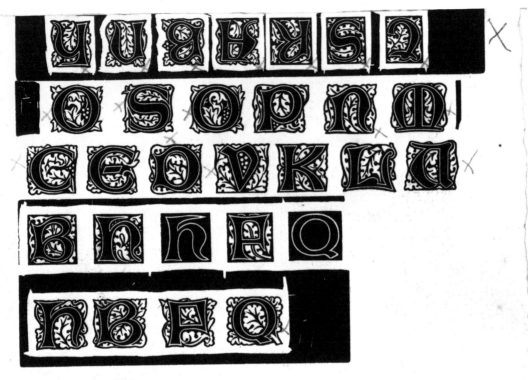

The white borders to the letters want all opening out especially those with a cross ×

Initials designed by William Morris for the Kelmscott Press

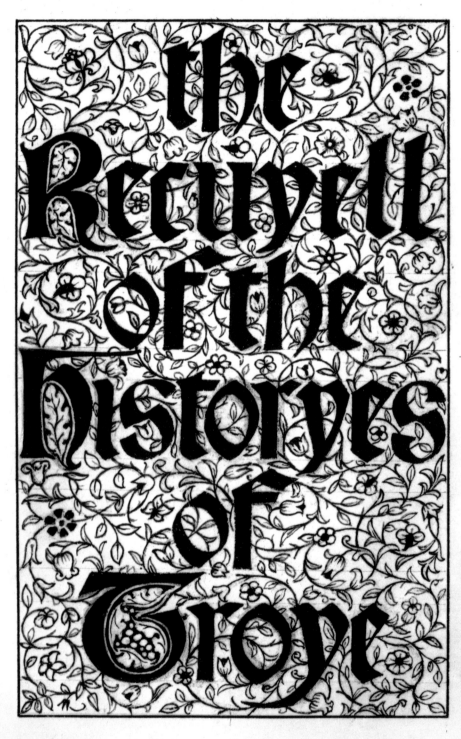

Original design by William Morris for the titlepage.

81в

LXXXV

The Story of Beowulf

And first of ~~his~~ the Kindred of Hrothgar.

WHAT! we of the Spear-Danes of yore days, so wabit
That we learned of the fair fame of Kings of the folk,
And the athelings a-faring in framing of valour.
Oft then Scyld the Sheaf-son from the hosts of the Scathers,
From kindreds a many the mead-settles tore
It was then the Earl feared them, sithence was he first
Found bare and all-lacking; so solace he bided,
Waxed under the welkin in worship to thrive,
~~Until~~ Until it was so that the roundabout sitters
All over the whale-road must hearken his will,
And yield him the tribute. A good king was that.

By whom then thereafter a son was begotten,
A youngling in garth, whom the great God sent thither
To foster the folk; and their crime-need he felt,
The load that lay on them while lordless they lived
For a long while and long. He therefore, the Life-lord,
The Wielder of glory, world's worship he gave him
Brim Beowulf waxed, and wide the weal upsprang
Of the offspring of Scyld in the parts of the Scede-land.
Such wise shall a youngling with wealth be a-working
With goodly fee-gifts toward the friends of his father,
That after in eld-days shall ever bide with him,
Fair fellows well-willing when wendeth the war-tide
Their lief lord a-serving. By praise deeds it shall be
That in each and all kindreds a man shall have thriving.

Then went his ways Scyld when the shapen while was,
All hardy to wend him to the lord and his warding:
Out then did they bear him to the side of the sea-flood,
The dear fellows of him, as he himself prayed them
While yet his word wielded the friend of the Scyldings

THE STORY OF BEOWULF
I. And first of the kindred of Hrothgar

WHAT! we of the Spear-Danes of yore days, so was it ⬥ That we learn'd of the fair fame of Kings of the folks ⬥ And the athelings a-faring in framing of valour ⬥ Oft then Scyld the Sheaf-son from the hosts of the scathers ⬥ From kindreds a many the mead-settles tore ⬥ It was then the earl fear'd them, sithence was he first ⬥ found bare and all-lacking; so solace he bided ⬥ Wax'd under the welkin in worship to thrive ⬥ Until it was so that the round-about sitters ⬥ All over the whale-road must hearken his will ⬥ And yield him the tribute. A good king was that.

BY whom then thereafter a son was begotten ⬥ A youngling in garth, whom the great God sent thither ⬥ To foster the folk; & their crime-need he felt ⬥ The load that lay on them while lordless they lived ⬥ For a long while and long. He therefore, the Life-lord ⬥ The Wielder of glory, world's worship he gave him:

LXXXVIII

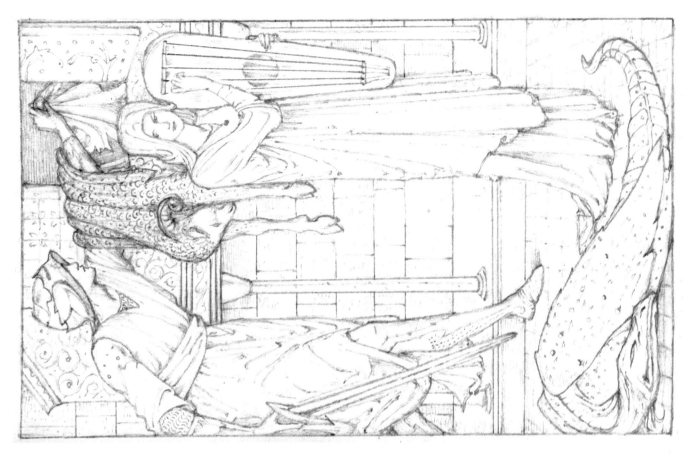

LXXXIX

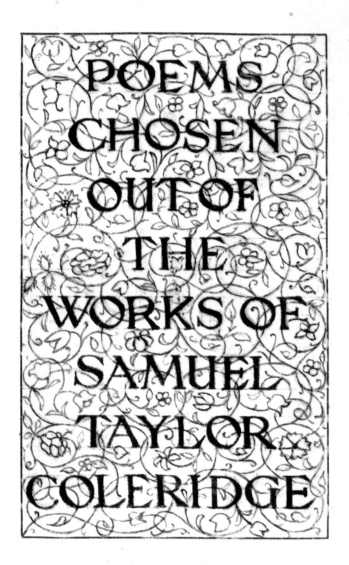

for Mr Keates Dec 17 1895

Original design by Morris for the title of this book

86 A

XCI

86a

86b

XCII

THE STORY OF SIGURD THE VOLSUNG AND THE FALL OF THE NIBLUNGS. BOOK I. SIGMUND

IN THIS BOOK IS TOLD OF THE EARLIER DAYS OF THE VOLSUNGS, AND OF SIGMUND THE FATHER OF SIGURD, AND OF HIS DEEDS, AND OF HOW HE DIED WHILE SIGURD WAS YET UNBORN IN HIS MOTHER'S WOMB.

I. Of the dwelling of King Volsung, & the wedding of Signy his daughter.

THERE WAS A DWELLING OF KINGS ERE THE WORLD WAS WAXEN OLD; DUKES WERE THE DOOR-WARDS THERE, & THE ROOFS WERE THATCHED WITH GOLD; EARLS WERE THE WRIGHTS THAT WROUGHT IT, AND SILVER NAILED ITS DOORS; EARLS' WIVES WERE THE WEAVING-WOMEN, QUEENS' DAUGHTERS STREWED ITS FLOORS, And the masters
of its song-craft were the mightiest men that cast
The sails of the storm of battle adown the bickering blast.
There dwelt men merry-hearted, and in hope exceeding great
Met the good days and the evil as they went the way of fate:
There the Gods were unforgotten, yea whiles they walked with men,
Though e'en in that world's beginning rose a murmur now and again
Of the midward time and the fading and the last of the latter days,
And the entering in of the terror, and the death of the People's Praise.

THUS was the dwelling of Volsung, the King of the Midworld's Mark,
As a rose in the winter season, a candle in the dark;
And as in all other matters 'twas all earthly houses' crown,
And the least of its wall-hung shields was a battle-world's renown,
So therein withal was a marvel and a glorious thing to see,
For amidst of its midmost hall-floor sprang up a mighty tree,
That reared its blessings roofward, and wreathed the roof-tree dear
With the glory of the summer and the garland of the year.
I know not how they called it ere Volsung changed his life,
But his dawning of fair promise, and his noontide of the strife,
His eve of the battle-reaping and the garnering of his fame,
Have bred us many a story and named us many a name;
And when men tell of Volsung, they call that war-duke's tree,
That crowned stem, the Branstock; and so was it told unto me.

SO there was the throne of Volsung beneath its blossoming bower,
But high o'er the roof-crest red it rose 'twixt tower and tower,
And therein were the wild hawks dwelling, abiding the dole of their lord;
And they wailed high over the wine, and laughed to the waking sword.

STILL were its boughs but for them, when lo on an even of May
Comes a man from Siggeir the King with a word for his mouth to say:
"All hail to thee King Volsung, from the King of the Goths I come:
He hath heard of thy sword victorious and thine abundant home;
He hath heard of thy sons in the battle, the fillers of Odin's Hall;
And a word hath the west-wind blown him (full fruitful be its fall!),

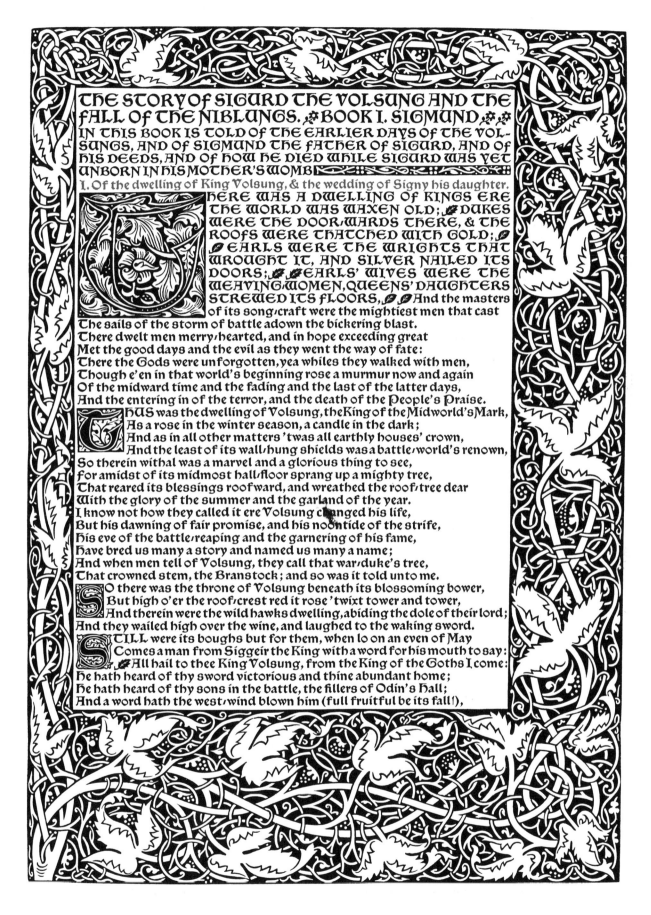

Chaucer's Works

1895						
				2578	3139	18	8
Decr 31	To Wages to date per Weekly Sheets				228	6	1
	,, do of Foreman & Reader (averaged)				62	9	6
	,, do Secretary do				52	4	10
	,, Paper used from Stock				82	3	8
	,, Vellum do				103	12	—
	,, Cost of Designs, Burne Jones £500						
	Others (includg Walker & Bowtall) 425.11.2				925	11	2
	,, Share of General Wages, not specifically apportionable				54	15	6
	,, do Paper & Vellums				73	6	4
	,, General Expenses & Deprn		223 11 3				
1896				479 9 3	4722	7	8
Mar 31	To Wages to date per Weekly Sheets				188	3	1
June 30	,, do				120	18	2
Octr 3	,, do of Foreman & Reader (averaged)				87	17	6
	,, do Secretary do				53	7	9
	,, Paper used from Stock				89	5	11
	,, Vellum do				152	—	—
	,, Cost of Designs, Burne Jones £200						
	Others (includg Walker & Bowtall) 255.9.—				455	9	—
	,, Editing fee to F. S. Ellis				150	—	—
	,, Binding Leighton 276.8.6. C Sanderson £61.10				337	18	6
	,, Share of General Wages, not specifically apportionable				62	18	3
				479 9 3	6420	5	10

And rente doun bothe wal, and sparre, and raf-
tur;
And to the ladies he restored agayn
The bones of here housbondes that were slayn,
To do exequies, as was tho the gyse.
But it were al to long for to devyse
The grete clamour and the waymentynge
Which that the ladies made at the brennynge
Of the bodyes, and the grete honour
That Theseus the noble conqueror
Doth to the ladies, whan they from him wente.
But schortly for to telle is myn entente.
Whan that this worthy duk, this Theseus,
Hath Creon slayn, and Thebes wonne thus,
Stille, in the feelde he took al night his reste,
And dide with al the contre as him leste.
To ransake in the cas of bodyes dede
Hem for to streepe of herneys and of wede,
The pilours dided businesse and cure,
After the bataile and discomfiture.
And so befil, that in the cas thei founde,
Thurgh girt with many a grevous blody
wounde,
Two yonge knightes liggyng by and by,
Both in oon armes, clad ful richely;
Of whiche two, Arcite hight that oon,
And as an aungel hevenly sche song.
The grete tour, that was so thikke and strong.
Which of the castel was the cheef dongeoun,
(Ther as this knightes weren in prisoun,
Of which I tolde yow, and telle schal)
Was evene joynyng to the gardeyn wal,
Ther as this Emels hadde hire pleyynge,
Bright was the sonne, and cleer that morwen-
ynge,
And Palamon, this woful prisoner,
As was his wone, by leve of his gayler
Was risen, and romed in a chambre on heigh,
In which he al the noble cite seigh,
And eek the gardeyn, ful of braunches grene,
Ther as the fresshe Emelye the scheene
Was in hire walk, and romed up and doun.
This sorweful prisoner, this Palamon,
Gooth in the chambre romyng to and fro,
And to himself compleynyng of his woo;
That he was born, ful ofte he seyd, alas!
And so byfel, by aventure or cas,
That thurgh a wyndow thikke and many a barre
Of iren greet and squar as eny sparre,
He cast his eyen upon Emelya,
And therwithal he bleynte and cryed, a!
And rente doun bothe wal, and sparre, and raf-
tur;
And to the ladies he restored agayn
The bones of here housbondes that were slayn,
To do exequies, as was tho the gyse.
But it were all to long for to devyse
The grete clamour and the waymentynge
Which that the ladies made at the brennynge
Of the bodyes, and the grete honour
That Theseus the noble conquerour
Doth to the ladies, whan they from him wente.
But schortly for to telle is myn entente.

And dide with al the contre as him leste.
To ransake in the cas of bodyes dede
Hem for to streepe of herneys and of wede,
The pilours diden businesse and cure,
After the bataile and discomfiture.
And so befil, that in the cas thei founde,
Thurgh girt with many a grevous bloydy
wounde,
Two yonge knightes liggyng by and by,
Both in oon armes, clad ful richely;
Of whiche two, Arcite hight that oon,
And as an aungel hevenly sche song.
The grete tour, that was so thikke and strong
Which of the castel was the cheef dongeoun,
(Ther as this knightes weren in prisoun,
Of which I tolde yow, and telle schal)
Was evene joynyng to the gardeyn wal,
Ther as this Emels hadde hire pleyynge,
Bright was the sonne, and cleer that morwe-
ynge,
And Palamon, this woful prisoner,
As was his wone, by leve of his gayler
Was risen, and romed in a chambre on heig
In which he al the noble cite seigh,
And eek the gardeyn, ful of braunches gren
Ther as the fresshe Emelye the scheene
Was in hire walk, and romed up and doun.
This sorweful prisoner, this Palamon,
Gooth in the chambre romyng to and fro,
And to himself compleynyng of his woo;
That he was born, ful ofte he seyd, alas!
And so byfel, by aventure or cas,
That thurgh a wyndow thikke and many a
Of iren greet and squar as eny sparre,
He cast his eyen upon Emelya,
And therwithal he bleynte and cryed, a!
And rente doun bothe wal, and sparre, and
tur;
And to the ladies he restored agayn
The bones of here housbondes that were s
To do exequies, as was tho the gyse.
But it were al to long for to devyse
The grete clamour and the waymentynge
Which that the ladies made at the brennyng
Of the bodyes, and the grete honour
That Theseus the noble conqueror
Doth to the ladies, whan they from him we
But schortly for to telle is myn entente.
Whan that this worthy duk, this Theseus,
Hath Creon slayn, and Thebes wonne thus
Stille, in the feelde he took al night his rest
And dide with al the contre as him leste.
To ransake in the cas of bodyes dede
Hem for to streepe of herneys and of wede
The pilours diden businesse and cure,
After the bataile and discomfiture.
And so befil, that in the cas thei founde,
Thurgh girt with many a grevous blody
wounde,
Two yonge knightes liggyng by and by,
Both in oon armes, clad ful richely;
Of whiche two, Arcite hight that oon,
And as an aungel hevenly sche song.

HERE BEGINNETH THE TALES OF CANTERBURY AND FIRST THE PROLOGUE THEREOF

What Aprille with hise shoures soote
The droghte of March hath perced to the roote,
And bathed every veyne in swich licour
Of which vertu engendred is the flour;
Whan Zephirus eek with his swete breeth
Inspired hath in every holt and heeth
The tendre croppes, and the yong sonne
Hath in the Ram his halfe cours yronne,
And smale foweles maken melodye
That slepen al the night with open eye,
So priketh hem Nature in hir corages,
Thanne longen folk to goon on pilgrimages,
And palmeres for to seken straunge strondes,
To ferne halwes, kowthe in sondry londes;
And specially, from every shires ende
Of Engelond, to Caunterbury they wende
The hooly blisful martir for to seke
That hem hath holpen whan that they were seeke.

Bifil that in that seson on a day,
In Southwerk at the Tabard as I lay,
Redy to wenden on my pilgrymage
To Caunterbury with ful devout corage,
At nyght were come in to that hostelrse
Wel nyne and twenty in a compaignye,
Of sondry folk, by aventure y falle

XCVIII

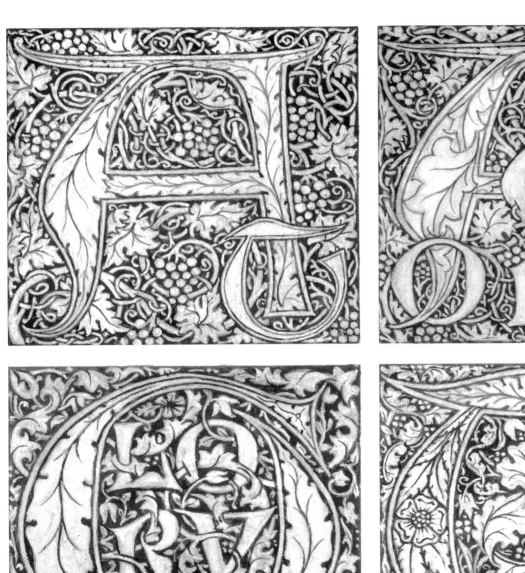

93D

And loveth him, the which that right for love
Upon a cros, our soules for to beye,
First starf, and roos, and sit in hevene above;
For he nil falsen no wight, dar I seye,
That wol his herte al hoolly on him leye.
And sin he best to love is, and most meke,
What nedeth feyned loves for to seke?

Lo here, of Payens corsed olde rytes,
Lo here, what alle hir goddes may availle;
Lo here, these wrecched worldes appetytes;
Lo here, the fyn and guerdon for travaille
Of Jove, Appollo, of Mars of swich rascaille!
Lo here, the forme of olde clerkes speche
In poetrye, if ye hir bokes seche.

O moral Gower, this book I directe

To thee, and to the philosophical Strode,
To vouchen sauf, ther nede is, to corecte,
Of your benignitees and zeles gode.
And to that sothfast Crist, that starf on rode,
With al myn herte of mercy ever I preye;
And to the Lord right thus I speke and seye:

Thou oon, and two, and three, eterne on lyve,
That regnest ay in three and two and oon,
Uncircumscript, and al mayst circumscryve,
Us from visible and invisible foon
Defende; and to thy mercy, everichoon,
So make us, Jesus, for thy grace digne,
For love of mayde and moder thyn benigne!
Amen.
Explicit Liber Troili et Criseydis.

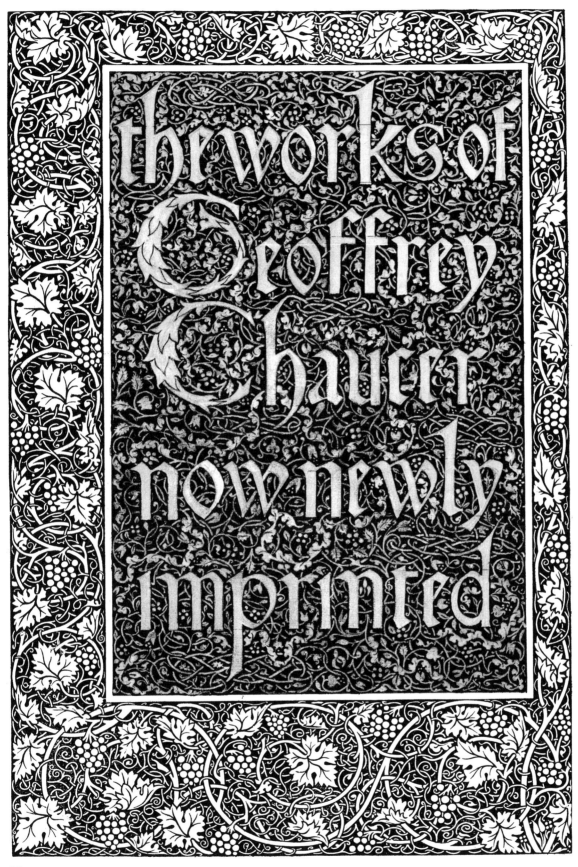

1 A B C
2 Complaint of Pity
3 Death of Dutchess
4 Complaint of Mars
5 Anelida and Arcite
6 Former Age

/1
/4 ½
/1
/1
12 ½

(13)
70
6
50
12
18
25
20
80

294
250

544

order
1 Canterbury Tales — 12
2 A.B.C. — ½
3 Complaint of Pitie — 1
4 Complaint of Mars }
5 ———— Venus } 3
6 Complaint of Anelyda — 4
7 Former age — ½
8 Good Conceil —
    a Prouerb
9 ~~A prayer to the Virgin~~
    Scogan
10 ~~& Proverb~~ } 1
11 Lines to Buxton
12 Gentelesse
13 Visage sans peinture } 1
14 Complent to his purse — 1
    Romance
15 ~~Parliament of Foules~~ 70
    Parliament of Faules
16 ~~Boethius~~ 6
17 Boethius — 50
18 Death of Blanche — 12
19 Astrolabie — 18
20 Legend of Good Women — 25
21 House of Fame — 20
22 Troilus — 80

95 в

93E

95 G

CVI

951

95 K

CVIII

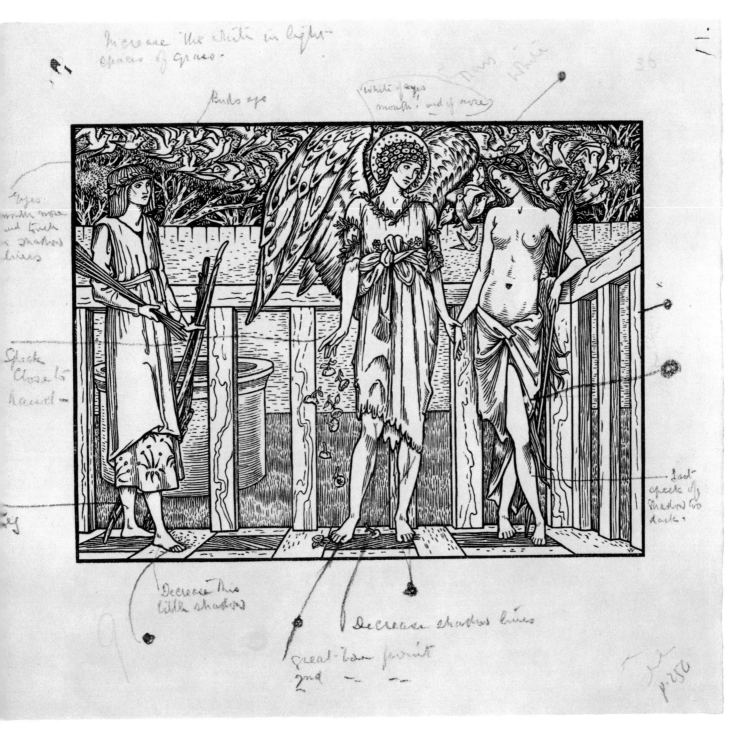

CX

100 c

Edward Burne-Jones
August 11th 1897.
The Grange. West Kensington.
b. Aug. 28 1833. d. June 18, 1898.

Designed the 87 illustrations

William Harcourt Hooper
Augt. 16th 1897
5, Hammersmith Terrace. W.
d. Feb 24 1912 aged 77

Engraved the illustrations on wood as well as many of the borders &c

Emery Walker   Aug. 16th 1897
3, Hammersmith Terrace W.
b. 2 April 1851

Gave constant advice and assistance

F. S. Ellis
The Red House
Cockington. Nov. 5. 1897
d. Feb 26 1901

Edited the text, usually followed that of Professor Skeat

Walter W. Skeat
2. Salisbury Villas, Cambridge
March 25, 1911.
b. Nov. 21 1835 d. Oct. 7 1912

R. Catterson-Smith.
16. Frederick Rd
Edgbaston - Birmingham -
24th May 1917.
b. 24 Feb 1853

Redrew Burne-Jones's pencil designs in ink. They were then photographed onto the woodblocks

Sydney C. Cockerell
3 Shaftesbury Rd Cambridge
May 24 1917 -
b. 16 July 1867

Was Secretary to the Kelmscott Press when this book was printed.

PRODUCED BY
THE STINEHOUR PRESS
AND
THE MERIDEN GRAVURE COMPANY

Date Due

APR 2 2 1989			
AUG 2 6 1991			
AUG 2 3 1991			
OCT 1 0 1999			
OCT 0 3 1999			
OCT 3 0 1999			
NOV 1 1999			
NOV 0 8 1998			

f Z232   .M87N4
Needham, Paul
   William Morris and the art of
the book.

ISSUED TO 273668

273668

HAMMERSMITH TERRACE

1 Doves Press 1900-1908
3 Emery Walker 1879-1903
3 Edward Johnston 1905-1912
7 Cobden-Sanderson 1897-1903
7 Emery Walker 1903-1933